NUCLEAR POWER

Nuclear Fission Reactors

James A. Mahaffey, Ph.D.

For Michael Thomas Fletcher

NUCLEAR FISSION REACTORS

Facts On File, Inc.
An imprint of Infobase Learning
132 West 31st Street
New York NY 10001

Library of Congress Cataloging-in-Publication Data
Mahaffey, James A.
Nuclear fission reactors / James A. Mahaffey.
p. cm.—(Nuclear power)
Includes bibliographical references and index.
ISBN 978-0-8160-7651-2 (alk. paper)
1. Nuclear reactors—Juvenile literature. 2. Nuclear fission—Juvenile literature. I. Title.
TK9148.M323 2011
621.48'3—dc22 2010053705

Facts On File books are available at special discounts when purchased in bulk quantities for businesses, associations, institutions, or sales promotions. Please call our Special Sales Department in New York at (212) 967-8800 or (800) 322-8755.

You can find Facts On File on the World Wide Web at http://www.infobaselearning.com

Text design by Annie O'Donnell
Composition by Julie Adams
Illustrations by Bobbi McCutcheon
Photo research by Suzanne M. Tibor
Cover printed by Bang Printing, Brainerd, Minn.
Book printed and bound by Bang Printing, Brainerd, Minn.
Date printed: October 2011
Printed in the United States of America

10 9 8 7 6 5 4 3 2 1

This book is printed on acid-free paper.

Contents

Preface

Nuclear Power is a multivolume set that explores the inner workings, history, science, global politics, future hopes, triumphs, and disasters of an industry that was, in a sense, born backward. Nuclear technology may be unique among the great technical achievements, in that its greatest moments of discovery and advancement were kept hidden from all except those most closely involved in the complex and sophisticated experimental work related to it. The public first became aware of nuclear energy at the end of World War II, when the United States brought the hostilities in the Pacific to an abrupt end by destroying two Japanese cities with atomic weapons. This was a practical demonstration of a newly developed source of intensely concentrated power. To have wiped out two cities with only two bombs was unique in human experience. The entire world was stunned by the implications, and the specter of nuclear annihilation has not entirely subsided in the 60 years since Hiroshima and Nagasaki.

The introduction of nuclear power was unusual in that it began with specialized explosives rather than small demonstrations of electrical-generating plants, for example. In any similar industry, this new, intriguing source of potential power would have been developed in academic and then industrial laboratories, first as a series of theories, then incremental experiments, graduating to small-scale demonstrations, and, finally, with financial support from some forward-looking industrial firms, an advantageous, alternate form of energy production having an established place in the industrial world. This was not the case for the nuclear industry. The relevant theories required too much effort in an area that was too risky for the usual industrial investment, and the full engagement and commitment of governments was necessary, with military implications for all developments. The future, which could be accurately predicted to involve nuclear power, arrived too soon, before humankind was convinced that renewable energy was needed. After many thousands of years of burning things as fuel, it was a hard habit to shake. Nuclear technology was never developed with public participation, and the atmosphere of secrecy and danger surrounding it eventually led to distrust and distortion. The nuclear power industry exists today, benefiting civilization with a respectable percentage of the total energy supply, despite the unusual lack of understanding and general knowledge among people who tap into it.

This set is designed to address the problems of public perception of nuclear power and to instill interest and arouse curiosity for this branch of technology. *The History of Nuclear Power,* the first volume in the set, explains how a full understanding of matter and energy developed as science emerged and developed. It was only logical that eventually an atomic theory of matter would emerge, and from that a nuclear theory of atoms would be elucidated. Once matter was understood, it was discovered that it could be destroyed and converted directly into energy. From there it was a downhill struggle to capture the energy and direct it to useful purposes.

Nuclear Accidents and Disasters, the second book in the set, concerns the long period of lessons learned in the emergent nuclear industry. It was a new way of doing things, and a great deal of learning by accident analysis was inevitable. These lessons were expensive but well learned, and the body of knowledge gained now results in one of the safest industries on Earth. *Radiation,* the third volume in the set, covers radiation, its long-term and short-term effects, and the ways that humankind is affected by and protected from it. One of the great public concerns about nuclear power is the collateral effect of radiation, and full knowledge of this will be essential for living in a world powered by nuclear means.

Nuclear Fission Reactors, the fourth book in this set, gives a detailed examination of a typical nuclear power plant of the type that now provides 20 percent of the electrical energy in the United States. *Fusion,* the fifth book, covers nuclear fusion, the power source of the universe. Fusion is often overlooked in discussions of nuclear power, but it has great potential as a long-term source of electrical energy. *The Future of Nuclear Power,* the final book in the set, surveys all that is possible in the world of nuclear technology, from spaceflights beyond the solar system to power systems that have the potential to light the Earth after the Sun has burned out.

At the Georgia Institute of Technology, I earned a bachelor of science degree in physics, a master of science, and a doctorate in nuclear engineering. I remained there for more than 30 years, gaining experience in scientific and engineering research in many fields of technology, including nuclear power. Sitting at the control console of a nuclear reactor, I have cold-started the fission process many times, run the reactor at power, and shut it down. Once, I stood atop a reactor core. I also stood on the bottom core plate of a reactor in construction, and on occasion I watched the eerie blue glow at the heart of a reactor running at full power. I did some time in a radiation suit, waved the Geiger counter probe, and spent many days and nights counting neutrons. As a student of nuclear technology, I bring a near-complete view of this, from theories to daily operation of a power plant. Notes and apparatus from my nuclear fusion research have been requested by and given to the National Museum of American History of the Smithsonian Institution. My friends, superiors, and competitors for research funds were people who served on the USS *Nautilus* nuclear submarine, those who assembled the early atomic bombs, and those who were there when nuclear power was born. I knew to listen to their tales.

The Nuclear Power set is written for those who are facing a growing world population with fewer resources and an increasingly fragile environment. A deep understanding of physics, mathematics, or the specialized vocabulary of nuclear technology is not necessary to read the books in this series and grasp what is going on in this important branch of science. It is hoped that you can understand the problems, meet the challenges, and be ready for the future with the information in these books. Each volume in the set includes an index, a chronology of important events, and a glossary of scientific terms. A list of books and Internet resources for further information provides the young reader with additional means to investigate every topic, as the study of nuclear technology expands to touch every aspect of the technical world.

Acknowledgments

I wish to thank Dr. Don S. Harmer, professor emeritus, Georgia Institute of Technology School of Physics, an old friend from the Old School who not only taught me much of what I know in the field of nuclear physics but conducted a thorough and constructive technical edit of the manuscript. I am also fortunate to know Dr. Douglas E. Wrege, a longtime friend and scholar with a Ph.D. in physics from the Georgia Institute of Technology who is also responsible for a large percentage of my formal education. He did a further technical editing of the material. A particularly close, eagle-eyed edit was given the manuscript by my Ph.D. thesis adviser, Dr. Monte V. Davis, whose specific expertise in the topics covered in this work was extremely useful. Dr. Davis's wife, Nancy, gave me the advantage of her expertise, read the manuscript, and saved me from innumerable misplaced commas and hyphenations. Special credits are due Frank K. Darmstadt, editor at Facts On File, Suzie Tibor, photograph researcher, and Bobbi McCutcheon, artist, who helped me at every step in making a beautiful book. The support and editing skills of my wife, Carolyn, were also essential. She held up the financial life of the household while I wrote, and she tried to make sure that everything was spelled correctly, all sentences were punctuated, and the narrative made sense to a nonscientist.

Introduction

In 1945, the use of a new, top-secret weapon, the atomic bomb, had finally ended Japanese aggression in the Pacific theater. The invention and application of this unknown technology had been the largest confidential scientific and industrial effort in history, and the enemy forces had never gotten wind of it. Days after the surrender of the Empire of Japan, the United States made an unprecedented concession to the world. Within two weeks of the war's end, the U.S. Army Corps of Engineers published the final report from the atomic bomb development project and made it available to anyone, recovering only the cost of printing. This hardbound book, *Atomic Energy for Military Purposes,* by Henry DeWolf Smyth, rocked the world to its foundation, as it revealed the methods by which the United States had solved the problems of power production by nuclear means. In the course of doing so with only four years of effort, the United States had progressed from an unimportant backwater of theoretical physics to the world's center for nuclear physics and technology.

The United States had entered World War II woefully underprepared, hauling howitzers using mules and arming soldiers with bolt-action rifles and "turtle-shell" helmets. Four years later, the nation had dropped atomic bombs from stratospheric bombers, finding targets with ground-imaging radar. At the end of the war, the United States gave away the secret to unlimited power that was discovered, nurtured, and advanced almost to industrial practicality at great expense. The country had accomplished this quickly as hostilities raged across Europe, Asia, and the entire Pacific Ocean.

Technologically, everything had changed, and the public was enticed with speculation concerning further development of this nuclear weapons technology. It quickly developed into a promise of a new, controllable power source. The first theoretical designs shown to the public, in such magazines as *Popular Science* and *Popular Mechanics,* were misleadingly simple and offered not a hint as to the dangers involved. The first nuclear *reactor* design presented to popular culture showed an irregular lump of *uranium* ore at the bottom of a steel water tank, heating the water to the boiling point. The steam was conducted through a pipe to a turbine used to turn an electrical generator. The power level of this simplistic reactor was controlled by a single piece of cadmium metal, looking like a ship's rudder suspended in the water.

Practical nuclear power generation has proven to be much more complicated than these first fanciful designs demonstrated. A modern *pressurized water reactor (PWR)* inside thick shielding walls looks as if 1,000 pipe fitters have competed to see who could make the most complicated tubing bends. Entire spreading rooms are used just to house thousands of electrical cables conducting instrument and actuation signals to and from the control room.

In the earliest days of nuclear reactor development, the control room consisted of a desk with a chart recorder pen sitting on it, scribing an inked line on a slowly moving roll of paper, and the control was a man holding the end of a rod protruding from the side of the reactor. The reactor control room has expanded greatly since then and now consists of row after row of tall instrument panels, each thickly populated with switches, lights, meters, digital readouts, graphic display screens, and alarm annunciators. Taken all together, a nuclear power plant is intimidating in its complexity, and it seems impossible to grasp its workings.

Nuclear Fission Reactors starts with an explanation of the nuclear *fission* process and how a controlled chain reaction can lead to a great deal of energy unlocked and allowed to escape from atomic nuclei. The second chapter details how energetic *neutrons* and fission fragments released in a fission reaction can be tamed and used as energy sources, and the third chapter reveals how this wild, seemingly unstoppable deluge of energy can be controlled and easily harnessed. Chapter 4 delves into the art of transferring heat from the source to the application of the heat, and it discusses how nuclear reactor cores avoid melting through the floor of the power plant building. The many practical designs for nuclear power plants that are now running are listed and discussed in chapter 5. In chapter 6, the important, nonnuclear components in a power plant are described, from the start switch on the control console to the generator at the end of the turbine shaft.

In chapter 7, the reader is led on a tour of a nuclear power plant, which most people have never visited. It is a tour of a General Electric *boiling water reactor* plant, where the author's expertise is centered. No two power plants are alike, but they have enough in common that the reader is exposed to the most important common characteristics of this branch of technology. Chapter 8 is a serious discussion of the questions that are legitimately asked concerning the large-scale use of nuclear power as a basis for electrical energy. The topics include air pollution, thermal pollution, the ultimate availability of uranium, *fuel reprocessing,* and finally the bottom-line cost of nuclear power and how it can emerge as an economical, competitive process as the world gradually changes. The final question about nuclear power, whether the waste products of nuclear fission can be dealt with safely, is covered in the conclusion.

Nuclear Fission Reactors will make sense of all this by breaking it down into small parts and explaining their purposes and functions. There are presently several subtypes of nuclear reactors being used as energy sources worldwide. The author's quarter century of work in nuclear research and engineering is fully exercised, making the interesting differences, quirks, and advantages of each variation clear to the reader. Although nuclear reactors have turned out to have more moving parts, more pipe runs, and more information-gathering systems than could have been imagined 70 years ago, the purpose of this book is to make this complicated technique understandable and to fill in the thin spots in our collective knowledge.

1 The Nuclear Fission Process

A modern nuclear power plant is basically a steam engine, using technology that was brought to full development in the 19th century. The working fluid is water, purified so as not to leave minerals sticking to the inside of pipes. As is the case in any steam plant, the water is brought to a boil by a heat source. The resulting vapor, or steam, is then used to convert the heat energy of the source into rotational energy, turning a generator shaft to make electricity. For this conversion, the highly energetic steam is directed to a turbine, through pipes. The turbine acts like a windmill, with the steam blowing through a collection of slanted vanes, tightly bound to a rotating wheel.

Steam-driven power plants are designed to get as much energy as possible out of the steam, so the vapor is directed through dozens of turbine stages, blowing through wheel after wheel, losing speed and energy as it gives them up to the vanes. Finally, when the steam's useful energy is spent, it is cooled and condensed back down into water. The water is pumped back into the boiling pot, and the cycle continues, in one endless loop, using the same water, over and over, to make high-pressure steam for the turbine wheels.

The electric generator, connected directly to the turbine by a common, horizontal shaft, is run at a precise speed, making the electrical current oscillate back and forth through the wires exactly 60 times per second, or at 60 hertz. The electricity exits the turbine building through three wires, into a large, outdoor matrix of circuit breakers and transformers, shifting the voltage level to tens of thousands of volts for conduction to faraway electrical consumers. Before the electricity shows up at the wall outlet in the home, it has been down-converted to 110 volts, still alternating direction at 60 hertz.

This description of electrical power generation fits 90 percent of the consumed electricity in the world. There are many details that can be added into the description, but this is fundamentally how it works. Most of the steam in these plants is made using burning coal as the heat source. Some is made by burning oil or natural gas, and a small amount is made using geothermal heat, or the heat from volcanic activity near the surface of the Earth. Alternative ways to turn the generator shaft are by falling water, or hydroelectric power, and a method of growing popularity is to use windmill power. Some power is made using the wave action of tides coming in and receding at the seashore, turning waterwheels as does the falling water in a hydroelectric dam.

Most of the world's electrical power, however, is still produced by steam, and 14 percent of this steam is produced not by burning coal, oil, or natural gas. It is produced by nuclear fission.

FISSIONABLE AND FISSILE NUCLIDES

Practical nuclear fission produces power because when a heavy atomic *nucleus* is cut into two major pieces, the mass of the pieces added together is less than the mass of the whole nucleus. This mass deficit is expressed as energy. This phenomenon was predicted by Albert Einstein's (1879–1955) mass-energy equivalence theory.

Technically, any atomic *nucleus* can be broken in two if it is hit hard enough with an incoming, high-speed particle, such as a fast-moving neutron. All *elements* except hydrogen are breakable. Hydrogen cannot be fissioned because its nucleus consists of only one particle, an unbreakable proton. To be fissioned, a nucleus must have at least two components. Most of the mass in the universe consists of unbreakable hydrogen *atoms.*

There are also tiny amounts of 91 other elements that compose the universe, and each of these elements has several variations, or *isotopes.* An element and its chemical properties are defined by the number of protons in its nucleus. Variations of an element contain different numbers of neutrons in the nucleus. While the neutrons do not contribute anything to the way an element combines with other elements, its melting point, or its electrical and magnetic characteristics, the neutron configuration in the nucleus determines all of an element's nuclear characteristics. These properties include an element's stability, or its tendency to self-destruct. The number of neutrons in the nucleus also determines whether or not an element can be fissioned the hard way, with a great deal of expended energy, or whether it can be fissioned the easy way, using a slow-moving neutron.

While many *nuclides* of the 92 naturally occurring elements may be *fissionable,* only a few are *fissile,* meaning that not only do they fission, but the fissioning process produces more energy than is required to achieve fission with an incoming neutron. The fission therefore produces net energy, most of which is the kinetic energy of the fissioned pieces, as if the nucleus has exploded. Moreover, a fissile nuclide must not only be able to break into two major fragments, but it must also produce more than one neutron in the breakup.

Nuclides, or variations of elements, in the actinide series of elements are generally able to fission productively, provided they have an odd number of neutrons in the nucleus. The actinide (actinoid) series is the group of characteristically *radioactive* elements in the periodic table of the elements above and including actinium. In this grouping of 14 elements, all but three are artificially produced and not naturally occurring.

A disc of pure uranium metal—normally silvery, uranium is highly reactive, and it turns dark gray and eventually black from oxidizing in the atmosphere. *(DOE photo)*

Of all the fissionable nuclides in the actinide series of elements, only four are defined as fissile and can actually be used as fuel in a nuclear reactor. They produce net power in fissioning, and they also produce more than enough free neutrons to continue the fission process. These special nuclides are uranium-233, *uranium-235,* which occurs naturally as a small component of mined uranium ore, *plutonium-239,* and plutonium-241. Uranium-233 is converted, or "bred," from naturally occurring thorium-232 by neutron capture. Thorium is the first element in the actinide series, and its 232 nuclide is often considered to be a legitimate nuclear reactor fuel. Technically it is not, as it is not fissile, but it converts so easily into fissile uranium-233 in the neutron-rich environment of a reactor core that it is considered a stand-alone nuclear fuel.

Uranium-238 is the major component of naturally occurring mined uranium, and it is an inert contaminant in power reactor fuel. However, as a uranium-fueled nuclear reactor produces power, it also produces plutonium-239 as a secondary process. Stray neurons have a probability of being captured by *uranium-238* and converting it to fissile plutonium-239. The plutonium can be extracted from spent uranium fuel and recycled back into the power production process as a fissile additive to fuel. There have also been power reactors built to run only on plutonium-239, specifically designed to breed more plutonium-239 than is used in power production by directing excess fission neutrons into sections of inert uranium-238.

Plutonium-241 is a smaller component of the production of plutonium-239. Plutonium-239, produced by neutron capture in uranium-238, can capture another stray neutron and become plutonium-240. *Plutonium-240* has an even number of neutrons and is therefore non-fissile. However, the plutonium-240 can capture yet another neutron and become fissile plutonium-241. It is usually a small component of plutonium fuel, but it does contribute to productive fission in a nuclear reactor.

ENERGY-DEPENDENT NUCLEAR REACTIONS

For these four fissile nuclides, fission by collision with a free neutron is not guaranteed. As is the case in all subatomic processes, such as nuclear decay, the process of fission is probabilistic, and the probability of fission is energy dependent. The energy or the speed of an incoming neutron can vary, from a high of 10 *million electron volts (MeV),* down to barely moving, or the speed of adjacent molecules jostling about because of their temperature. This thermal energy at room temperature, or 72°F (20°C), is only 0.025 *electron units (eV).* A thermal neutron is traveling at 7,218 feet per second (2,200 m/s). A high-speed neutron, at a moving energy of 10 MeV, is traveling at close to the speed of light. At high speeds, the neutron interaction probability with a nucleus is generally low.

The probability of neutron interaction with a nucleus, whether the process is simple scattering or bouncing off, absorption of the neutron by the nucleus, or fission of the nucleus, is expressed as an effective cross section of the nucleus, in square centimeters. Individual atomic nuclei are rather small, and a fairly large cross section is 10^{-24} square centimeters. This is in the ballpark of the actual cross-sectional area of a very large nucleus, such as the proton-neutron cluster at the center of a uranium atom, but it can vary widely, depending on the process probability that it represents. Early in the development of nuclear science, in the 1940s, someone commented that a nucleus with a measured neutron interaction probability of 10^{-24} square centimeters was "as big as a barn," and, from that remark, nuclear cross sections have been expressed in barns ever since. A nuclear cross section of 1.0×10^{-24} square centimeters is 1.0 barn.

With neutrons traveling at thermal speed, examples of neutron interaction cross sections for various elements range widely. The non-fissioning neutron absorption cross section for oxygen is only 0.0002 barns, but for gadolinium it is 46,000 barns. The fission cross section for uranium-238 is zero, but for uranium-235 is 577 barns. This unusually large neutron fission probability for slow neutrons in uranium-235 makes it possible to have a self-sustaining *chain reaction* of fissions. This favorable probability of fission overcomes many obstacles to achieving controlled power production.

THE CHAIN REACTION

The chain reaction is essential to power production by neutron-induced fission. In a solid mass of uranium, containing a percentage of fissile uranium-235, an incoming neutron from an external source will very likely come into contact with a suitable nucleus and cause a fission, releasing more than 200 MeV of energy. Among the debris from this incident are likely two neutrons. Each of the two neutrons blasts off in its own direction, and there is a probability that one of them will find another uranium-235 nucleus in the uranium mass and cause another fission. The reactions are sequential, as it takes a finite length of time for the neutron to travel the distance separating the two atoms. A fission can thus produce another identical reaction in another uranium-235 atom. This phenomenon of one fission causing another fission after a slight delay is called a chain reaction. The chain reactions can continue, making a continuous series of fissions, as long as the probability of fission is favorable.

There are many ways to achieve unfavorable results. The most important factor is the size of the mass of uranium. If the size were infinite, then every neutron born of fission would be likely to hit another uranium nucleus, but in a small mass, there is a good probability of a neutron escaping into space without ever being stopped by a nucleus. A neutron produced near the surface of a mass will be just as likely to fly off the surface as to plunge into the middle of the mass, as the flight direction is entirely random. This deterrent to fission is called leakage. The slightest impurity in the uranium, particularly an impurity with a very large nonproductive absorption cross section, will capture neutrons and deny them the opportunity to produce fission.

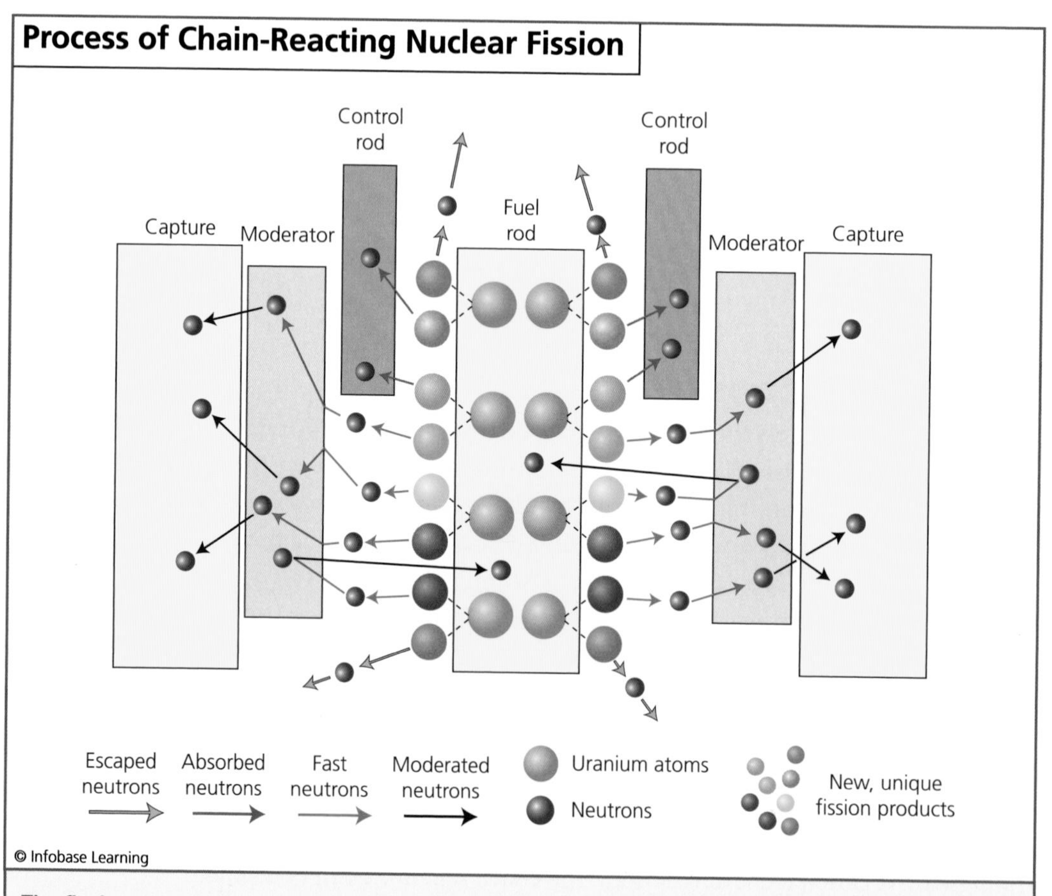

The fission process in a nuclear reactor. The debris of a uranium fission event consists of two smaller nuclei plus two neutrons, set free. A neutron can be absorbed by a control rod or escape the machine or be slowed down to fission speed by the intervening moderator material, leading to another fission.

Various measures are used to ensure an infinite chain reaction in a nuclear reactor. They include using an adequate-sized reactor core, *enrichment* of the percentage of fissile material in the core over non-fissile material, reduction of the amount of nonproductive contaminants, and even consideration of the shape of the reactor core. A practical power reactor core can be as small as a garbage can or as big as a two-story house, depending on the purity percentage of uranium-235 in the fuel. The largest reactors use natural fuel, which is only 0.7 percent uranium-235. The smallest reactors, such as those used for submarine power, use uranium artificially improved to 50 percent uranium-235.

The most advantageous shape of a reactor core is a perfect sphere, as a sphere has the optimum volume-to-surface-area factor. The second-best shape is a cylinder with the proportions of a tomato can. Cans used to hold portions of food are scientifically designed to use the least amount of metal, making the most possible volume. They therefore have the best volume-to-surface-area factor for a cylindrical shape.

THE DELICATE BALANCE OF NEUTRON GAIN AND LOSS

It seems easy to start a power-producing chain reaction in fissile fuel. All that is necessary is a mass of fuel big enough to sustain the reaction and one neutron to start things off. One neutron produces one fission, but the one fission produces two neutrons. Those two neutrons produce two more fissions, which produce four neutrons. Those four fissions produce eight neutrons, and so on, as the number of fissions multiplies as rapidly as neutrons can travel the space between atoms. It is as simple as it sounds. Each fission produces 200 MeV of heat, and the heat from fissions rapidly adds up to enough energy to boil water and spin a turbine.

The only problem with this process is stopping it before the fission machine, the nuclear reactor, melts into a runny, glowing river of metal. To operate in steady state, in which the reactor produces a fixed, predetermined rate of heat, the neutron production must be precisely controlled. At a modest power rate of a few watts, a nuclear reactor produces hundreds of trillions of neutrons per second. The precise number depends on the size, the shape, and the power level of the reactor, but for a given reactor at an exact power level the number of neutrons produced per second must be precise. To achieve this condition, the production rate of neutrons in the reactor must absolutely match the neutron loss rate. Neutrons are produced by fission and are lost by unproductive leakage from the surface

THE MOUSETRAP ANALOGY

Dr. Heinz Haber (1913–90) was a German physicist who immigrated to the United States after World War II. He first worked in aviation medicine research at Randolph Air Force Base, later becoming an associate physicist at the University of California, Los Angeles, in 1952. From his duties at the university, he was channeled into a consulting job at Walt Disney Studios. The Walt Disney Company was working on an exciting new television program on the ABC network named *Disneyland.*

Disneyland offered cartoons, edits of Disney films, and a miniseries about Davey Crockett. In 1956, Disney featured a special film on the benefits of nuclear power called *Our Friend the Atom.* Heinz Haber, who was specially skilled in making complex scientific issues understandable, was tasked with writing and designing the show. He wound up starring in it. With a combination of animated sequences and live-action filming, Haber did an excellent job of explaining the concepts of nuclear energy, which were then unknown to just about everyone. His clear depictions of the atom, the nucleus, and then nuclear fission were capped with an ingenious demonstration of the fission chain reaction. He did it with a room full of steel-spring mousetraps, all set to go off.

Each mousetrap represented a uranium atom, steady and yet ready to snap at the slightest touch, with two Ping-Pong balls nestled in the spring. The Ping-Pong balls

of the reactor core, unproductive absorption, and by productive absorption, or fission. For every neutron that is born in fission, one must be lost.

The state of absolute balance between production and loss, at which the reactor power level is steady and fixed, is called *criticality.* A critical reactor is one that has achieved this state of absolute balance. A subcritical reactor is losing more neutrons than it is making, and the power level is therefore dropping. A supercritical reactor is making more neutrons than it loses, and therefore the power level is climbing. The degree of supercriticality or subcriticality determines the speed with which the power level changes.

Recall, it requires only one neutron to start a chain reaction. If a reactor is losing 100 trillion neutrons per second but is making 100 trillion plus one neutrons per second, then it is supercritical and its power level will rise exponentially. In an exponential power rise, as the power increases, represented the two neutrons that would be produced in a uranium fission. The walls of the room were mirrored, to give the impression that the floor of traps extended into the far distance. Haber picked up a Ping-Pong ball and explained that it represented a neutron. He then tossed the ball into the room. It hit the trip lever on a mousetrap, which went off, sending its two balls flying wildly. Those two balls immediately hit two more mousetraps, which sent their four balls flying. In the blink of an eye, the entire room was awash in flying Ping-Pong balls, with mousetraps going off with a loud, prolonged bang. It was highly dramatic, and it made the principle of chained fission perfectly clear. With no constraints over the speed of the balls or the multiplication of the number of balls, the reaction was totally out of control, and it obviously represented an atomic bomb explosion, or uncontrolled nuclear fission. As an encore, he then showed the Ping-Pong ball explosion in slow motion. At reduced speed, the mousetraps were visible jumping, spinning, and cartwheeling all over the room in a cloud of airborne balls.

The Disney film explaining nuclear power, which originally aired in 1957, holds up well with age and remains an excellent introduction to the advanced principles of nuclear energy. At the time the film was made, nuclear power, and the threat it represented during the cold war, was on everyone's minds. *Our Friend the Atom* is available on YouTube in five segments, beginning with this clip that may be viewed here: http://www.youtube.com/watch?v=ZcdRQkJulAU.

the rate of increase goes up. On the other hand, if a reactor is making 100 trillion neutrons per second and losing 100 trillion plus one neutrons per second, then it is subcritical and is losing power exponentially.

The next chapters will clarify how it is possible to easily control such a small perturbation of such an enormous production rate. The next chapter will detail the process of slowing down the dynamics of the fission reaction, and the chapter after that will disclose the methods of controlling a billion watts of power with the touch of a button.

2 How to Make Energy with Nuclear Fission

The energy released by a single uranium fission is about 210 MeV, and, of that, 200 MeV are recoverable or are able to be converted to useful power. The energy from this one fission is enough to make a particle of dust jump. This may not seem a lot of energy, but the number of fissions per second in a nuclear reactor is enormous. Fissions at the rate of 3.1×10^{10} per second produce one watt of usable power. The typical commercial power reactor makes 1 billion watts of power.

To make power at the rate of a megawatt, or enough power to supply 1,000 houses with 1,000 watts of lighting, a reactor need only consume one gram of uranium-235 per day.

It is possible to build a nuclear reactor the size of a cantaloupe, as a simple sphere of pure plutonium-239, a fissile nuclide. Any practical power plant is much larger, requiring a multistory building to contain the reactor core and its containment vessel. However, the power level that can be achieved by any nuclear reactor, regardless of its size, has no theoretical limit. The reactor the size of a cantaloupe could theoretically supply power to the entire world.

There is, however, a practical problem. A reactor is a physical object, made of solid material, and it has a melting point, as does every other solid object in the universe. The problem is not making power, the problem is removing power from the source so that heat will not build up, increase its temperature, and melt the reactor. The power level of any reactor design is therefore limited not by the physics of fission but by the limitations of the process of removing heat from the reactor. The fuel in this theoretical cantaloupe-sized reactor would also burn out in a flash, as the rate of energy consumption in the world is about 15 terawatts, or 15 trillion watts. The fuel would be totally consumed in 52 seconds, assuming that the chain reaction could somehow be sustained as the amount of fissile uranium dwindled away rapidly.

The practical power production rate of a nuclear power reactor is about 1,000 megawatts, or 1 billion watts. Practical reactor fuel is not enriched to nearly pure uranium-235 but is inert uranium-238 containing a small concentration of fissile uranium-235, about 3 percent. Such a low-grade fuel must be coaxed into a sustained chain reaction by optimizing the factors that make fission possible. There is also the problem of removing 1 billion watts from the reactor core and transferring it to the electrical generator. These engineering problems are solved in one step, using a feature called the *moderator.*

THE THERMAL REACTOR AND THE USE OF A MODERATOR

In the first few decades of nuclear power development following World War II, many possible reactor designs were tried in an effort to find the best way to build a practical nuclear power system. To be practical, a system must be safe to operate and economical to run. It must have a long lifetime and be easy to maintain. These simple criteria whittled down the large set of possible reactor plant designs to a few solid candidates. Most of the nuclear power plants in current use around the world use thermal reactors.

Thermal refers to the *thermal speed* of the neutrons used to cause fissions. Neutrons of any available speed can cause fissions, but neutrons slowed down to the relatively slow speed of common molecules are called thermal neutrons. At this speed, the probability of fission in uranium is optimized, and a lesser concentration of the fissile uranium-235 may be used in the fuel.

Neutrons materialize in a fission event at fairly high energy and speed, up to 10 MeV, and to reach thermal speed they must be slowed down to 0.025 eV, which is a factor of more than 10 million. The way to slow neutrons down is to collide them with a slow-moving object of similar size or mass. The slowing-down process is a purely classical operation and does not involve any quantum mechanics. In fact, slowing down a neutron is done the same way a billiards player slows down a cue ball. Hit a stationary billiard ball with a well-shot cue ball, and the cue ball will come to a stop as the stationary ball shoots away, taking on the speed of the cue ball. Neutrons work the same way.

As it turns out, a proton weighs about the same as a neutron, so protons are used to slow neutrons by collision. The nucleus of a hydrogen atom is a single, naked proton—perfect for stopping a high-speed neutron in its tracks, or at least slowing it to the speed of the hydrogen atom. Hydrogen is easy to include in the structure of a nuclear reactor core. It is a major

A photo of a reactor core running at full power. The water being used as a moderator between fuel rods glows a bright blue, due to Cerenkov light. This glow is due to beta particles from fission product decay passing through the water at speeds greater than the speed of light in water. *(NASA)*

component of water (H_2O)—two atoms of hydrogen and one atom of oxygen, chemically bound together. An optimum way to slow neutrons to fission speed is to immerse uranium fuel, formed into thin rods, in water. Neutrons set free in one fuel rod will be slowed down as they traverse the short distance to an adjacent fuel rod. It takes 18 collisions, on average, to slow a neutron down to thermal energy.

There are many other materials that can be used as a moderator, such as solid blocks of graphite or beryllium, but an additional advantage to using water as a moderator is that it is also an excellent coolant. Water in the liquid phase is easily pumped into the bottom of the reactor core. It absorbs the heat of fission as it travels upward through the core, and the hot water is extracted at the top. The constantly moving coolant keeps the fuel from melting, and it acts as a transfer medium for the heat, conducting it out of the core and to adjacent machinery to convert the heat to electricity. The dual role of water, as both a moderator to make fission possible and as a coolant to stabilize the core temperature, is highly advantageous.

Ordinary water, however, has one disadvantage. The hydrogen can, on rare occasion, absorb a neutron, becoming heavy hydrogen, or *deuterium*. The deuterium nucleus weighs twice that of a simple hydrogen nucleus, as it contains one proton and one neutron. The occasional loss of neutrons in hydrogen capture is just enough to make it impossible to run a water-moderated reactor using natural uranium, which is only 0.72 percent fissile uranium-235. A water-moderated reactor must use slightly enriched fuel, having an artificially enhanced uranium-235 concentration.

There is a class of successful power reactors, notably the *Canada Deuterium Uranium (CANDU)* reactors, that can run on natural uranium. The CANDU is water moderated, but the water is special *heavy water,* or water made with deuterium instead of simple hydrogen. A given volume of deuterium is literally twice as heavy as a similar volume of simple hydrogen, and deuterium's ability to slow neutrons is less than that of hydrogen. There is not the clean, total transfer of momentum when a billiard ball, or a neutron, hits an object with twice its weight, but the slowdown can still occur. It takes more collisions, 25, to completely lose the initial speed in heavy water than it does in ordinary water.

Deuterium, however, never absorbs a neutron. It already has a neutron. The advantage of no neutron capture makes a deuterium-moderated

(continues on page 16)

THE ADVANTAGES AND DISADVANTAGES OF GRAPHITE

At the earliest point of nuclear reactor development in 1940, graphite was seen as an attractive material for a neutron moderator. Graphite is a crystalline form of carbon. The carbon atom is much heavier than a hydrogen atom, and, for this reason, a neutron must hit 114 carbon atoms to lose enough energy to slow to thermal speed. A reactor with graphite as the moderator must therefore be significantly larger than a reactor with water as the moderator.

However, purified graphite does not have a tendency to absorb any neutrons, so none are taken out of the fission process, and this fact alone makes it an ideal moderator. It is, in fact, ideal to the point that natural uranium can be used to produce a critical, self-sustaining chain reaction in a graphite-moderated reactor. The process of artificially enhancing the uranium-235 content in mined uranium is difficult and expensive, requiring a very large industrial effort. Rather than spend years first building a uranium enrichment capability, scientists decided to use graphite moderation in the first generation of experimental power plant reactors. Water-moderated reactors came along later, after a uranium enrichment industry was established and a need for compact reactors developed.

Parallel efforts started in Germany to develop nuclear power technology at the same time as in the United States, right before the start of World War II. The German scientists came to the same conclusion as the American scientists, that a nonabsorptive moderator was necessary for the first reactor experiments, and graphite was a prime choice. The graphite for a reactor must be pure of any contaminants, as any nongraphite material in the moderator was likely to absorb neutrons and kill the effect of the graphite. German industry was capable of making graphite, but it was contaminated with boron, which rendered it useless. The Germans turned to heavy water as a moderator, and production problems plagued the German experiments for the duration of the war. A working reactor was never assembled.

In the United States at the same time, a production facility the size of Rhode Island was built in Washington State, having several large plutonium production reactors, all moderated with chemically pure synthetic graphite. After the war, Great Britain and the Soviet Union built their own versions of the high-power plutonium production reactors and even electrical power reactors using graphite as an economical moderator. It allowed them to run their reactors on low-grade natural uranium fuel instead of the expensive enriched fuel. These reactors would eventually reveal severe problems with the use of graphite as a moderator.

FP-1 fuel, or uranium oxide pseudospheres, nestled in a block of graphite moderator, as used in the first nuclear reactor on December 2, 1942 *(Argonne National Laboratory, courtesy AIP Emilio Segrè Visual Archives)*

The first problem was called the Wigner effect. Neutrons colliding with carbon atoms in the graphite crystal matrix would slow down, but in doing so they were likely to knock carbon atoms askew in the matrix, disturbing the crystal structure. This reaction would literally cause the crystal to change shape, distorting the mechanical integrity of the reactor, which was basically a huge pile of graphite bricks neatly arranged in a cylindrical shape. The energy expended to move these atoms was stored in the graphite, and there was danger of it all releasing at once. The graphite had to be annealed periodically, taken to an unusually high temperature for a controlled energy release. In 1957, a graphite-moderated plutonium production reactor at the Windscale facility in Cumbria, England, caught fire during an annealing operation, destroying the reactor and contaminating a section of the country with *fission products.* Graphite, as it turns out, can burn fiercely in air, and the fire is difficult to extinguish. Water thrown on a graphite fire just makes it burn hotter.

In Soviet Ukraine in Chernobyl in 1986, a large graphite-moderated power reactor, a Soviet-designed reaktor bolshoy moshchnosty kanalny (RBMK), a high-power channel reactor, caught fire and exploded, contaminating much of western Europe with pulverized fuel and highly radioactive fission products. No graphite-moderated reactors have been designed or built since then.

(continues from page 13)

reactor able to use a lower grade fuel, and it still operates as a coolant for the reactor. Advanced CANDU reactor designs use heavy water as the moderator and circulate ordinary water as the coolant, keeping the two substances separated.

In ordinary water, as found in lakes, rivers, and oceans, about one in every 3,200 hydrogen atoms is the heavy, deuterium variation. Effort is required to separate out the rare deuterium-loaded water molecules using electrolysis, distillation, or a chemical process, and this makes it expensive. The current price of 2.2 pounds (1 kg) of heavy water is between $600 and $700. The cost of the heavy water moderator in a CANDU reactor is 20 percent of the cost of the entire plant.

THE FAST REACTOR AND FUEL BREEDING

Almost every nuclear power reactor in the world uses the advantages of the high fission probability of thermal neutrons. These reactors use ordinary water moderators, heavy water moderators, and a few obsolete examples still use graphite to slow down the neutrons. There is an alternative reactor design that uses fast neutrons, fissioning uranium at nearly the speed with which they emerge from fission, with an energy of more than 1.0 MeV. These rare, exotic units are fast reactors. They are also known as breeders, or the two names can be combined into fast breeders.

For reasons to be discussed, the fast breeders are temperamental, almost impractical, reactors to build and run as electrical power production units, but they do have an overriding and attractive feature. A fast *breeder reactor,* running and producing power, makes more fuel than it burns. An entire power economy could be built using breeder reactors, in which there would never be a lack of fuel.

A typical breeder burns plutonium-239 as its fissile fuel. Plutonium-239 fissions readily using high-speed neutrons. It is a synthesized element and is made without non-fissile impurities, as natural uranium has in abundance. A breeder reactor, using fast neutrons, requires no moderator material for efficient fission. In fact, a fast breeder reactor core is specifically designed to waste fission neutrons, having them leak out the sides of the core as a significant fraction of the neutrons produced. These wasted neutrons are then captured by a surrounding blanket of uranium-238, which converts into plutonium-239 upon neutron capture.

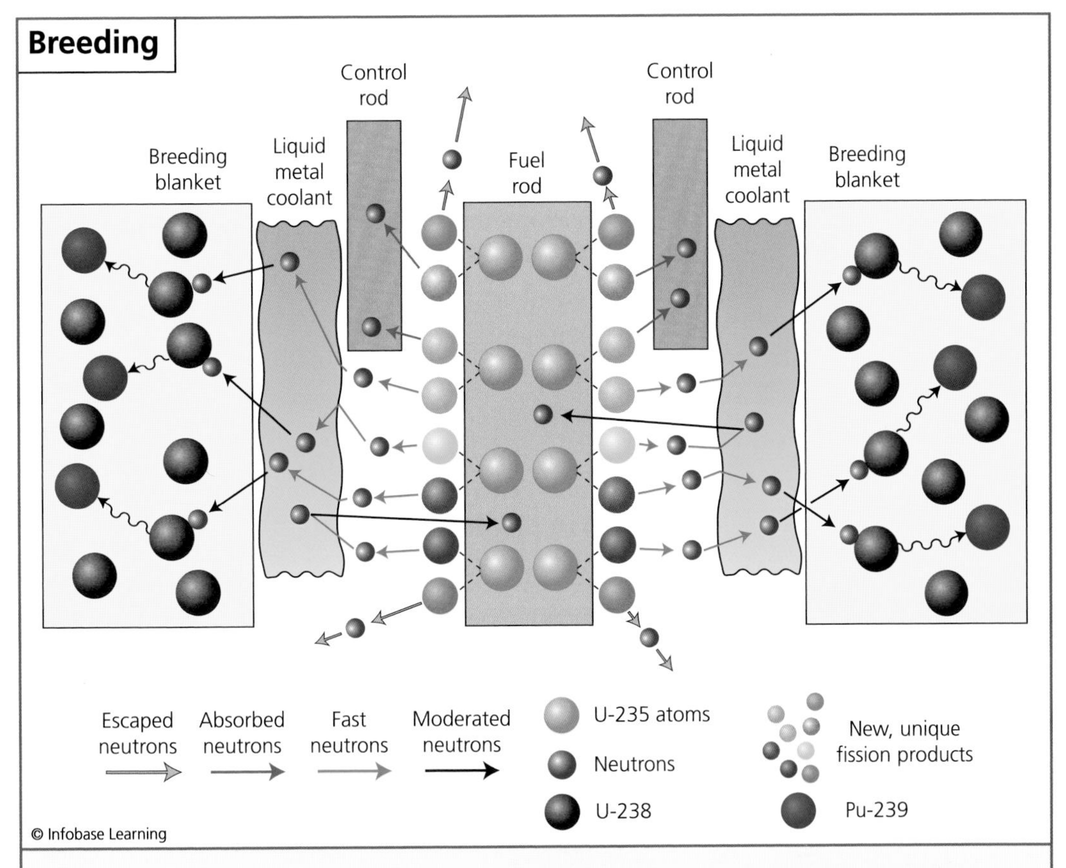

In the plutonium breeding process, there is no moderator. The coolant is liquid metal, and fast neutrons from fission cause further fissions, escape the machine, are captured by control rods, or make it to the breeding blanket, where they transmute uranium-238 into plutonium-239.

For each plutonium-239 atom that is fissioned in the power production process, more than one plutonium-239 atom is made in the uranium blanket. The breeder reactor thus makes more burnable fuel than it consumes.

The uranium-238, being more than 99 percent of all mined uranium and otherwise worthless, is plentiful and inexpensive. There is enough of it to run the world's electrical power requirements for hundreds of thousands of years in a breeder economy by using it as breeder stock in fast reactors. Uranium-238, or depleted uranium, is the waste left over in the uranium enrichment production for the nuclear power industry.

About once a year the working fast breeder reactor is shut down, so that the expended fuel can be removed from the core, and the uranium-238 blanket is removed. The newly bred plutonium-239 is then chemically separated from the blanket, formed into fuel elements, and a portion of it is reloaded into the core. The fuel cycle then repeats.

This is not a speculative futuristic technology. In fact, the first electrical power ever produced by a nuclear reactor was produced by a fast breeder reactor, the experimental breeder reactor (EBR-1) near Arco, Idaho. At 1:50 P.M. on December 20, 1951, the electrical generator connected to EBR-1 was switched on, powering four 200-watt lightbulbs. This experimental facility eventually powered its entire building until it was finally shut down permanently in 1964. It had proven the concept of using a fast breeder reactor as a power source that could produce its own fuel for perpetuity.

Bigger, more powerful breeders were built. EBR-1 was rebuilt as EBR-2. In 1966, the Fermi 1 fast breeder reactor first produced 94 megawatts

The Superphénix breeder reactor at Creys-Malville, France, on August 28, 1990. It is shut down for changing the internal sodium filters, and the men walking atop the reactor give a sense of the size of this machine. *(© Roger Ressmeyer/CORBIS)*

of electricity for the Detroit Edison Power Company in Monroe County, Michigan. Since then, breeder reactor technology has been studied, and power-production breeders have been built in India, France, the United Kingdom, Russia, and Japan.

The fast breeder reactor would seem an ideal source of power. It makes no exhaust gases of any kind, and it replenishes its own fuel. There is, however, a drawback to the use of fast reactor technology. Its coolant must be pure metal, liquefied so that it may be circulated through the hot reactor core.

The reason for this requirement is contrary to the needs of the thermal neutron reactor. In the thermal reactor, all neutrons must be slowed to the advantageous slow speed. In a fast breeder reactor, the neutrons cannot be slowed down, because uranium-238 breeds plutonium-239 efficiently using only high-speed neutrons. Water therefore cannot be used as a coolant, as it would also act as a moderator. The plutonium-239 in its pure, non-diluted form fissions readily at the high neutron speed.

The specifications for the liquid metal coolant limit the number of available materials. It must melt and run freely at temperatures significantly below the melting point of the plutonium-239 fuel and of metal structures in the reactor vessel, pumps, and piping. Mercury would seem a good choice, but unfortunately it captures neutrons voraciously and discourages fission. The metal sodium has no large neutron capture cross section, and it melts at 208°F (98°C), below the boiling point of water at 212°F (100°C). Potassium metal has an even more favorable melting point, at 145°F (63°C), and it also has no great tendency to absorb neutrons.

Liquid sodium has proven to be a good, non-moderating coolant for fast reactors. In many applications, it has been used alone, and in some a mixture of sodium and potassium called NaK is used. Pure sodium and NaK suffer from one large disadvantage over water as a reactor coolant. If water should leak from a reactor's complex cooling system, the worst that it can do is possibly carry dissolved fuel or fission products to the leak site. Water is easily cleaned up and is easily replaced. Leaking sodium or NaK, on the other hand, is chemically dangerous. It reacts explosively with water or even with water vapor in the air, and the product of the reaction is sodium or potassium hydroxide. These compounds are extremely corrosive and will quickly dissolve aluminum reactor components or injure exposed skin. Another problem with these liquid metals is that at room temperature they are not liquid. They are solid, so if a reactor is shut down and cooled off for maintenance, the entire core and cooling system are encased in solid metal. The metal is opaque, and the condition of fuel assemblies and the inside of the core tank are not visible for quick inspection.

In a water moderator, fast neutrons convert their energy to heat in the coolant by being stopped by collisions with the light hydrogen or deuterium atoms. The neutrons lose energy in the collision, and the water gains energy. This is not the case with liquid metal coolant. All of the heat is generated in the fuel or in the breeding blanket of the fast reactor. Cooling is accomplished by conduction, with the hot fuel conducting heat to the relatively cool liquid metal. The fuel must be in constant contact with the coolant, and a momentary lapse of coolant flow allows a fuel assembly to overheat, melt, and wreck the carefully designed reactor core. Many of the early liquid metal–cooled reactors, including EBR-1, the sodium reactor experiment, and Fermi-1, experienced core meltdowns due to interrupted coolant flow.

Although the fast breeder reactor is an intriguing concept, engineering problems due to the liquid metal cooling have held back a total commitment to this type of power production. Because of the finite supply of naturally occurring fissile isotopes on Earth, a plutonium-239 breeder reactor economy may be civilization's hope of maintaining nuclear electrical power in the far future.

3 Controlling Nuclear Fission

In a nuclear power reactor, the fission process is kept at a steady level in which absolutely every neutron that is produced is balanced by a neutron that is lost. If more are produced than are lost, the reaction rate increases exponentially until damage to the reactor core shuts down the fission process. If fewer neutrons are produced than are lost, the reaction rate falls exponentially, also shutting down the fission process.

Much research and development has gone into the art of controlling the fission process in a nuclear reactor, keeping the fission process at precisely the critical level, or, at the touch of a button, moving the rate of power production up or down to a specific, predetermined level. This seemingly impossible task is not as difficult as it might seem, due to fortunate characteristics of how thermal nuclear fission occurs in reactor fuel.

THE DYNAMICS OF NUCLEAR FISSION

The rate of nuclear fission, or power level of a reactor, and maintenance of the perfectly critical level of fission are easily controlled because of a naturally occurring delay between each fission event and the beginning of the next fission. The delay is not very long, but it is sufficient to take the sensitivity out of attempts to control the fission. If fissions followed fissions instantly in a nuclear reactor, then the controls would be extremely sensitive, with the slightest change in the reactor configuration sending the power level off on a dangerous excursion. Instead, a change in the reactor is not instantaneous. It takes a finite length of time for a reactor to respond to any change, and this fact alone makes fission easy to control. A control error can be corrected before it has a chance to affect the rate of fission. Most reactor control, in fact, consists of constantly correcting slight errors, overshoots, or undershoots to maintain a reactor's power level at a constant level.

Slight errors in power-level maintenance are physically unavoidable because of the random, probabilistic nature of fission. Fission is random in all its aspects, starting with the probability of a fission occurring under ideal conditions, the direction in which neutrons are ejected from a fissioned nucleus, the number of neutrons produced in a given fission, and ending with the chance of a fission neutron intercepting a uranium-235 nucleus. The average number of neutrons produced in a uranium-235 fission is 2.43. Another way of expressing this probability is to say that out of every 100 fissions, 243 neutrons are produced. It is impossible to predict precisely how many neutrons are produced in a given fission. All of these tiny, probabilistic errors mean that a large nuclear reactor is randomly bumping from slight supercriticality to slight subcriticality all the time, with the time average condition being absolute criticality. This absolute criticality is always drifting slowly out of range, but it can be easily corrected by detecting a change in the power level and adjusting the controls. The built-in fission delay makes such corrections as easy as holding a car in its lane while driving on the expressway.

The fission delay may be broken into three components: the neutron slowing-down time, the neutron diffusion time, and the delayed neutron emission.

In ordinary water, it takes an average time of 7.1×10^{-6} seconds, or 7.1 microseconds, for neutrons to slow down to thermal speed from their high-speed separation from fissioned uranium-235. In heavy water, it takes almost 10 times that long, or 50 microseconds. In graphite, with its nuclei heavier than those of heavy water, it takes even longer, at 140 microseconds. The heavier nuclei require multiple hits to drain the energy out of traveling neutrons, and the slowing-down time is therefore longer. Microseconds may not seem a long time, but the three delay effects add up, and they result in a non-instantaneous response to controls.

The second delay effect, diffusion time, is a measure of how long a neutron must wander around in the moderator before it interacts with a uranium-235 nucleus. The fuel is always interspersed in the moderator, making it available for fission with thermalized neutrons. At thermal speed, the neutrons have no particular direction of motion. They just float around in the moderator, changing direction with every bump into a moderator nucleus. This action is referred to technically as the drunkard's walk, or diffusion. In ordinary water, the diffusion time is, on average, 240 microseconds. In heavy water, the drunkard's walk is 60,000 microseconds, and in graphite it is 16,000 microseconds.

In nuclear fission, the fissile nucleus breaks into two ragged pieces, and most likely two free neutrons are included in the fission debris. However, these free neutrons are not necessarily released immediately. An eventually free neutron can be bound up in one of the fission fragments, which are two lesser elements created by the breakup of the heavy uranium or plutonium nucleus. These two fragments are always radioactive, as the newly created elements are invariably built with a great excess of neutrons. These newly formed nuclides are unstable and *decay* to a stable configuration of protons and neutrons. A fraction of these *radioactive decays* eject neutrons, and these events occur over a wide range of probabilities. Any neutron that results from a fission-fragment decay can result in another fission, but the release of this neutron into the fission process depends entirely on the *half-life* of its parent nuclide.

The half-life of a nuclide is the time it takes for half of its quantity to decay. A freshly produced supply of 100 atoms of bromine-87, which is a common product of fission, would decay to 50 atoms of bromine-87 in 55.6 seconds.

The ejection of a new neutron from fission debris is not necessarily a direct action. The decay of bromine-87 does not eject a neutron, but the product of its beta-minus decay is krypton-87. Krypton-87 has 6 MeV of excess energy, and this is enough to eject a loosely bound neutron in its nucleus. The result is a stable krypton-86 atom plus one neutron.

The half-lives of possible delayed-neutron sources vary from 0.23 to 56 seconds. This separation between the fission cause and the fission effect is the major component of the neutron-delay effect. The fraction of fission neutrons that are delayed in uranium-235 fission is 0.65 percent, or an average of 0.0153 neutrons are delayed per fission. In a plutonium-239 reactor, 0.20 percent of the fission neutrons are delayed, or an average of 0.0061 neutrons per fission.

This delayed fraction of the neutrons produced in fission is an important feature. It slows down the response of a fissioning mass of uranium or plutonium to small changes in the system configuration, and it thus allows a very smooth, easy control of a nuclear reactor by mechanical means.

NEUTRON POISONS

There are many ways to control a nuclear reactor. All ways involve varying the number of fission neutrons that are allowed to participate in the process. The most common way to control a reactor is through the use of neutron poisons. These controls absorb neutrons parasitically, without causing or allowing any nuclear fission. If a neutron poison is formed into a rod or a blade, this structure can be inserted or removed from the core of the reactor, causing the total number of neutrons present at a given time to raise or lower. This action causes the rate of fission to increase in a supercritical reaction or to decrease in a subcritical reaction. With the controls positioned carefully in the reactor core, the fission is kept at the point of exact criticality. The reactor can be shut completely down and held in a safe condition by inserting all of multiple controls fully into the core. With a maximum amount of neutron poison interspersed with the fuel and the moderator, a reactor can produce no power and fission reactions cannot be self-sustaining.

A nuclear reactor must be built with excess reactivity. It must have enough fissile fuel configured in a way that is advantageous for fission for it to be supercritical, for two reasons. First, it must be capable of supercriticality because when running and producing power it will immediately lose fissile fuel and its ability to maintain fission. There must be enough excess fuel in a reactor to compensate for the loss of fissile fuel by burn-up. In the case of a

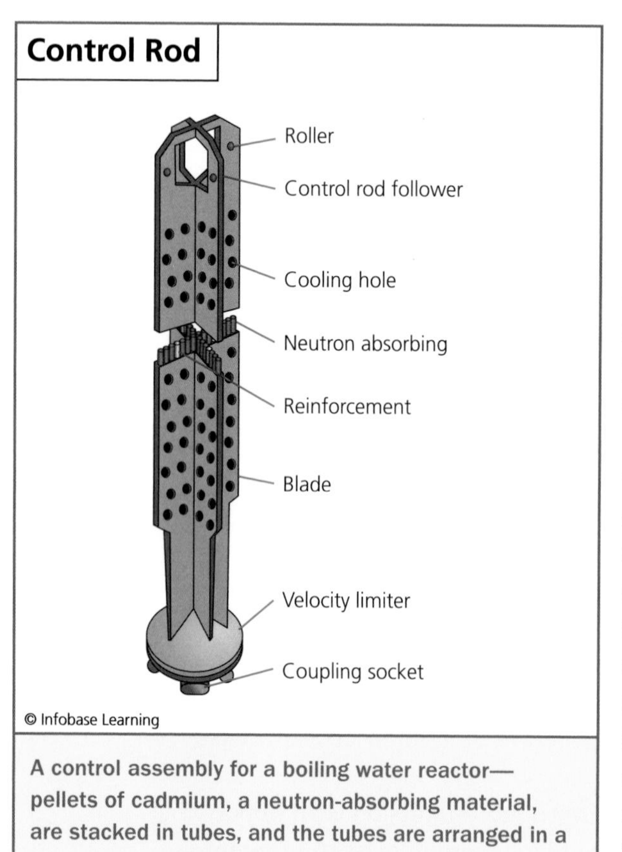

A control assembly for a boiling water reactor—pellets of cadmium, a neutron-absorbing material, are stacked in tubes, and the tubes are arranged in a cruciform configuration in a long metal shell.

A view of the bottom of a boiling water reactor vessel at the Kashiwazaki-Kariwa Nuclear Power Plant in Japan, showing the cluster of hydraulic drive mechanisms for the control rods *(Issei Kato/Reuters/Landov)*

large power plant, at least enough excess reactivity is built in so that the reactor can run at full power for at least a year. Neutron poison controls keep the excess fuel in check, so that even though it is in the reactor core it is not fissioning. As fuel burns, the controls are slowly withdrawn over the course of a year or years, adding to the effective fuel load.

A second reason for the excess reactivity is that to increase the power level, a reactor must be allowed supercriticality at a carefully controlled rate of rise. When a reactor is started up, from a condition of zero power

FISSION THAT CONTROLS ITSELF

All power reactors have active controls to precisely set the power level. A power reactor, loading electricity into a large, complex network of electricity consumers, must be adjusted constantly as the power demand waxes and wanes. There are several designs of reactor making power in the world. One of these basic designs, the boiling water reactor, is unique in that it controls itself as the power demand changes, with no external mechanisms or human attention necessary.

In the early days of power reactor development, immediately after World War II, a nuclear unit that was both cooled and moderated by water was seen as an attractive concept. One fluid would act to both slow down the neutrons and to transfer heat from the hot reactor to the power-generating equipment. The water, however, would have to be kept under high pressure to keep it from boiling in the reactor core. It was predicted that boiling would lead to instability, as the bubbles of steam produced by boiling would have neither moderating nor cooling abilities. The chaotic production of steam bubbles was not seen as appropriate in an orderly nuclear reactor core in which neutron production rates were kept constant.

In 1952, Samuel Untermyer II (1912–2001) made a prediction concerning the behavior of a reactor with a boiling moderator. He argued that the chaos created by forming random bubbles in the moderator would not contribute to the naturally chaotic behavior of nuclear fission and that the bubbles would actually have a benefit. The hotter a reactor core becomes, the more bubbles form. As bubbles form, the density of the moderator goes down. This degrades the action of the moderator, so the reactor goes slightly subcritical and the power level starts dropping. As the power level drops, the core temperature drops with it, and as the temperature drops the water stops boiling. The bubbles go away. The moderator regains its density, and the shutdown, it is put in a condition of supercriticality in which the fission is allowed to increase slowly. The neutron poison controls are withdrawn from the core, allowing more fission than is necessary to achieve self-sustained fission. When the desired power level, or rate of fission, is achieved, the controls are eased back slightly, bring the reaction down to just barely a critical level. Small vernier controls are then used to constantly adjust the power level and keep it stable as the reactor generates constant power.

power level then begins to rise. It is a closed loop, with beneficial consequences. The formation/destruction of steam bubbles in the core maintains a set power level without the motion of metallic neutron poisons.

Untermyer won a U.S. government contract to test his theories with physical experiments, and he directed the construction of an experimental boiling water reactor at the National Reactor Testing Station near Arco, Idaho. The first experiment, *BORAX-I,* successfully demonstrated the safety characteristics of this water-moderated reactor. Untermyer went on to build BORAX-II through BORAX-V over the next 20 years, demonstrating the full capabilities and advantages of this design using larger and more powerful power plants.

The major safety advantage of the boiling water reactor is that if the fission were to lose control, increasing in power level, all the water would boil away, leaving fuel without any moderator. Without moderator, the reactor would be unable to sustain fission, and it would be completely shut down without the motion of any controls. This is not true of graphite-moderated reactors, as graphite cannot boil away, nor is it true of any liquid metal reactors, which have no moderator action.

Today there are 35 operating boiling water power reactors in the United States, 19 operating in Japan, two in Finland, five in Denmark, two in India, two in Mexico, two in Spain, seven in Sweden, and two in Switzerland. The boiling water reactor is considered by many engineers to be the most inherently safe design currently in use.

This belief was tested on March 11, 2011, when a magnitude 9.0 earthquake followed by a tsunami wrecked all six boiling water reactors (BWRs) at the Fukushima I Nuclear Power Plant in Japan. Although these units were damaged beyond repair, there were no uncontrolled nuclear reactions and no breakage of the reactor vessels. The BWRs behaved under extreme conditions as they were designed to, but the quake and tidal wave were beyond engineering predictions for maximum stress.

There are many choices and many configurations for neutron poison materials in reactor controls. Materials are chosen for their ability to parasitically absorb neutrons, their mechanical strength, melting point, and cost. A common control material is cadmium. A component of naturally mined cadmium is the nuclide cadmium-133, which is 12.3 percent of all cadmium. It has a very large thermal neutron absorption cross section of 20,000 barns. Gadolinium-157 is also used, with an even larger absorption cross section of 240,000 barns.

Probably the best control material for a water-moderated reactor is the rare, expensive element hafnium. All of four naturally occurring nuclides of hafnium have large neutron absorption cross sections, with the largest being 380 barns. When a nucleus in a control material absorbs a neutron, it loses the ability to absorb neutrons and becomes inert to the process. When the 12.3 percent of cadmium-133 has absorbed enough neutrons, the ability to serve as a control is ended. In hafnium, 100 percent of the metal is a neutron absorber, so it remains effective for longer than other naturally occurring metals. Hafnium is similar chemically to zirconium, which is most often used for reactor fuel cladding and internal metal structures in a reactor core. Zirconium is resistant to corrosion in hot water, and it has a high melting point of 4,051°F (2,233°C).

The metal that is used as a neutron poison in a reactor is formed into shapes that can nest between stacks of fuel sitting in moderator material. The fuel, moderator, and controls are arranged closely together in a reactor design, with each material having its intended role in making nuclear power. The fuel provides the fission; the moderator brings fission neutrons down to the right speed; and the controls soak up just enough neutrons to make the reaction barely self-sustaining.

Depending on the type of power reactor, controls can be run from the top of the core, the bottom, or one side, but in all cases the controls must be capable of motion. Put the control deeper into the core, and the reaction rate slows. Pull it out, and the reaction rate increases. In the earliest experimental reactors of the early 1940s, controls were pushed and pulled horizontally, out the sides of graphite-moderated reactors. As large power-producing reactors were designed, controls became heavier and human access to the sides of a reactor became limited. Control motion was changed to geared electrical motors, allowing positioning by remote control. Instead of laboring at the side of a reactor, an operator could press a button in a control room to move a *control rod* in or out of the core.

Within days of the first sustained nuclear fission, in Chicago on December 2, 1942, an autorod was devised. It was a vernier control, shaped as a long rod of cadmium, with a reversible electric motor driver. The reactor power level was monitored by a neutron-counter instrument attached to the reactor, and the output of this counter was connected to the electric circuit of the drive motor. By pulling the rod out when the power level was dropping and pushing it back in when the power was increasing, the autorod kept this prototype reactor automatically at a constant fission rate. All the human reactor operator had to do was sit back and watch it work at constantly adjusting the vernier rod in and out of the core.

NONPOISONOUS CONTROLS

There are other ways to achieve full control of a nuclear reactor without using insertable neutron poisons. While a water-moderated reactor can achieve good safety control by generating steam bubbles, full control means bringing a reactor from a state of cold inactivity to a state of hot-running power production. The controls must also command sufficient reserve reactivity to overcome negative reaction influences. For example, neutron poisons can be created by fission products. These newly created nuclides will build up in the fuel as the reactor operates. At least one bit of fission debris, xenon-135, has the neutron absorption power to shut a reactor down if it lacks sufficient control range.

The simplest way to achieve this full range of control is to divide a reactor into two parts. Each part individually is subcritical, but when assembled together, the complete reactor is supercritical. Criticality is achieved by adjusting the proximity of the two pieces. This method has been used for experimental fast reactors running at room temperature, but it is not practical for large power reactors.

Another alternate control mechanism is to vary the reflection of neutrons back into the core as they naturally leak out. Normally, any fission neutrons that manage to find their way to the outer boundaries of the reactor are lost, as they contact the air and are not directed back into the reactor. However, it is possible to build a variable neutron reflector from a material such as beryllium. The degree of reflectivity is controlled by the way the reflectors are pointed. Neutrons that are redirected back into the core add to the fission activity and bias the reactor toward supercriticality. An example of a power reactor built with this unusual control scheme was the SNAP-10A, built in 1965 and operating at 30 kilowatts. It was a satellite-based power station, generating power experimentally in low Earth orbit.

For the wide range required for power reactor control, neutron poisons are used exclusively. Poisons such as boric acid can be mixed with water and used for an emergency shutdown measure. Dumping this borated water into a reactor needing a quick shutdown strangles any fission activity, while it also provides extra cooling water in an emergency. Boron is also used as a burnable poison in some reactor fuel.

A load of borated fuel gives a reactor an extended quantity of uranium that is guaranteed not to fission until the boron has been deactivated. This allows more fissile uranium-235 to be loaded into the reactor at one time without increasing the reactivity, and it extends the lifetime of the core to several years. The boron captures neutrons and takes them out of the fission process, but in doing so the boron loses its ability to absorb more neutrons, and it burns up. By the time the boron has burned up, the non-borated fuel in the reactor has burned up from fission. The newly exposed unfissioned uranium-235 then takes over to generate power.

BIO-SHIELDING

Controlling nuclear fission to maintain a constant power level at criticality is a straightforward process. Aside from the reactor core fission rate, one more aspect of nuclear power must be controlled. Nuclear fission creates intense *radioactivity* across the entire spectrum of *radiation* types and energies. The act of fission itself creates radiation, and the products or debris of the fission process are neutron-heavy nuclides. Each nuclide will go through several stages of nuclear decay, emitting radiation as it tends toward becoming a stable, nonradioactive nuclide with balanced numbers of neutrons and protons. A static fixed control must be in place to protect the power plant personnel and the public from this radiation.

Any mass of material will stop radiation. Air in the atmosphere protects people from heavy cosmic ray radiation constantly streaming down overhead. The heavier or more dense the material is, the more efficient it is at attenuating or completely extinguishing radiation. Material thickness is also a factor. Materials commonly used for bio-shielding are lead, concrete, steel, earth, and water.

The first barrier to fission radiation is the water in the reactor tank or vessel, used as a moderator and a coolant. A column of water 15 feet (4.6 m) deep will protect people from the most energetic gamma-ray emissions. A

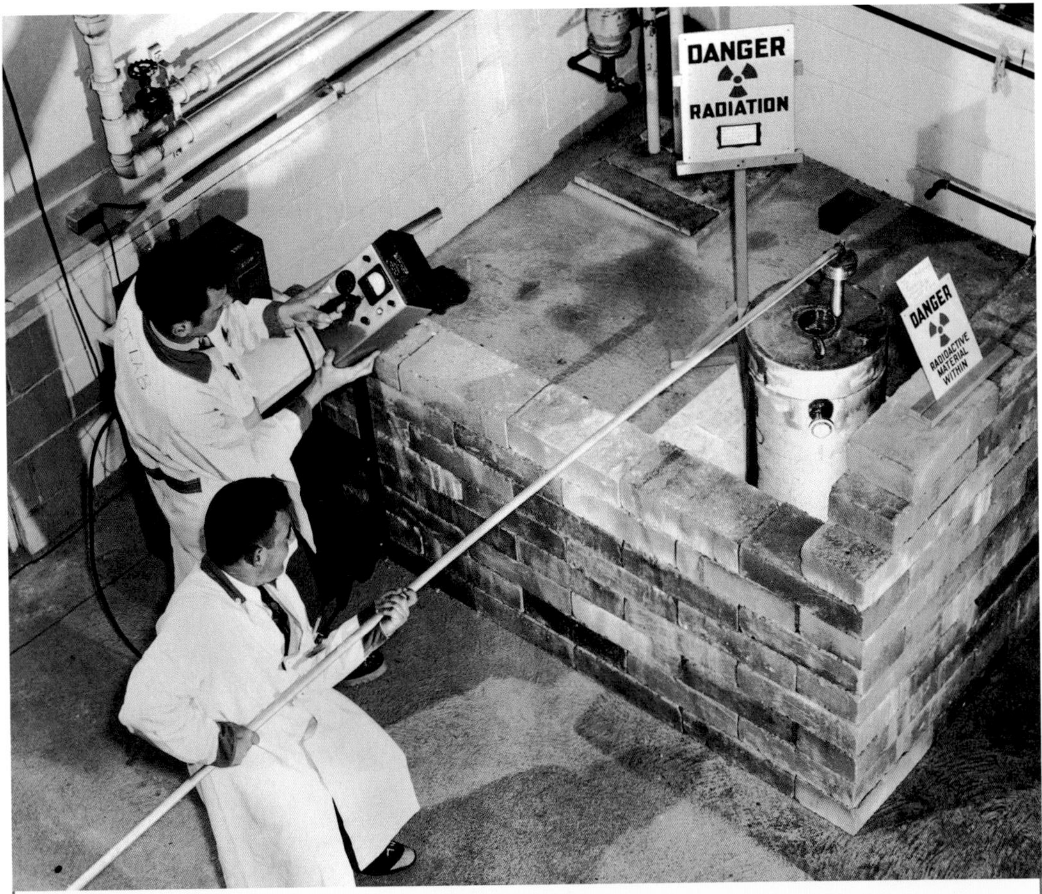

A generous supply of lead bricks between the radiation source and a human being is always a good shield. Here a technician removes the plug from a lead canister as his associate monitors the radiation with an instrument. *(© Bettmann/CORBIS)*

few inches of lead will have the same effect. The reactor tank, or vessel, itself is made of steel more than eight inches (20 cm) thick and built to withstand at least 2,000 pounds per square inch (14,000 kp) of steam pressure.

The thick steel reactor vessel is nestled in a steel containment structure, intended to prevent a worst-case breach of the reactor vessel from spreading radioactive material into the surrounding countryside. These structures are typically built of steel one inch (2.5 cm) thick, or its equivalent in another construction material such as concrete. This secondary enclosure acts as further shielding against radiation exposure to plant workers.

Special shielding is required for neutrons. Neutrons are neither electromagnetic radiation nor charged particles, and simple mass shielding does not stop them. Neutrons, in fact, will pass readily through thick lead shielding, which will stop the effects of any other radiation. Neutrons that have escaped the fission reactor core, however, are thermalized examples, and they can be stopped with a layer of a material with a high thermal neutron-absorption cross section. A prime example is boron. Boric acid, containing boron, can be mixed with concrete, paraffin, or water to provide a perfect bio-shield against stray neutrons.

It is thus possible to safely control the fission rate and the radiation emissions from a nuclear fission reactor. Reactors respond smoothly and predictably to varying degrees of neutron poisoning, and radiation can be squelched with the materials that normally go into the construction of an electrical power plant. The next challenge to producing nuclear power is to cool the reactor. Denied cooling, it will overheat and even melt.

4 Transferring Heat and Keeping Things from Melting

A nuclear reactor is incapable of suffering a nuclear explosion as if it were an atomic bomb or a thermonuclear weapon. An atomic bomb is an entirely different device, built to experience fast, hypercritical, prompt fission. A reactor is built for thermal, critical, delayed fission. At its worst, a nuclear reactor can suddenly overheat the water in the reactor vessel and experience a steam explosion. Such an occurrence is very serious, but it does not lead to the type of destruction wrought by a nuclear weapon. Nearly as bad as a steam explosion is a core meltdown, caused by a lack of coolant on the fuel as it is generating power. A meltdown destroys the reactor core, and secondary effects of such an accident can render a power plant unusable.

These destructive accidents are to be avoided at all costs, and for this reason much effort goes into the design and construction of the reactor cooling system. The cooling system in a power plant actually has two functions. It keeps the reactor core below its melting point, which is around 5,100°F (2,800°C), and it moves heat from within the reactor out to the power conversion machinery. The conversion turns raw heat into electricity.

Many schemes have been tried for power reactor cooling. Exotic coolants made of organic compounds, molten salt, lithium-7, and even mercury have been tried, but for practical power generation the coolant materials have settled into two categories, liquid and gas.

LIQUID COOLING

The most common material used for liquid cooling in a nuclear power reactor is ordinary water. Water is a well-known substance, having been used as a coolant and an energy-transfer medium in power production machinery since the 18th century. In the age of steam power, which was about 200 years, there were many innovations and refinements in the techniques of pumping and controlling heated water. Water is inexpensive, and it serves the double use of moderator and coolant. Water has low viscosity and, compared to other coolant liquids and gases, requires less pumping effort. Its thermodynamic properties, such as its ability to absorb heat, are well studied and are favorable for use in high-power nuclear applications. Water is the most popular coolant in use in nuclear reactors worldwide. Of the more than 400 nuclear power reactors presently operating, more than 90 percent are water cooled.

Water will flash into steam at temperatures and pressures that are compatible with normal building materials, such as carbon steel and stainless steel. This may seem an obvious qualification, but some coolants have been tried with questionable compatibilities. The use of liquid metals as coolant, for example, can introduce problems that do not exist in any water-cooling system. When a liquid metal system has cooled off, the metal has solidified into hard material that is opaque, it captures and holds all moving parts in the system, and it catches fire when exposed to air. In a water system, the pipes corrode, but in a liquid metal system, the coolant corrodes. Liquid metal, however, has excellent heat-conducting characteristics and is used whenever neutron-moderating effects are not desired.

A disadvantage to water cooling using ordinary water is that it tends to absorb neutrons, and this effect makes it impossible to run a water-moderated reactor on natural uranium fuel. The fuel must be enriched in uranium-235, and this is an expensive option. Water tends to corrode metal, and this can reduce the useful life of pipes, valves, and pumps. Water also decomposes into hydrogen and oxygen under heavy radiation bombardment, which is unavoidable in a nuclear reactor operating at high power.

The problem of neutron absorption can be solved by using heavy water, or water containing the heavy isotope of hydrogen. Its disadvantage is that it is expensive, but there are presently 48 CANDU reactors in the world operating with heavy water in the primary cooling loop.

The problem of decomposition under radiation, or radiolysis, is solved by recombining the free oxygen and hydrogen back into water using a catalytic converter, somewhat similar to the catalyst units used in gasoline-powered automobiles.

GAS COOLING

In the early days of power reactor design following the military developments of World War II, various gases were considered for use as coolants. Graphite was the most common neutron moderator in use at the time, and it made some sense to cool the solid mass of the reactor by blowing gas through it. Water-cooled systems were open-looped, depending on a nearby river for heat removal, and these watercourses were not always available. There was no need for additional moderating effect in a graphite reactor, and the gas was simply used to carry away heat.

An obvious gas to use for reactor cooling is air, which could be blown through horizontal holes drilled in the large graphite pile. Pellets of natural uranium fuel were distributed throughout the block of graphite. This simple scheme of reactor cooling had a serious drawback, in that the graphite could catch fire using the free oxygen in the air. More inert gases were employed after this problem resulted in a disastrous reactor fire at the Windscale facility in England in 1957. An attractive alternative was helium, as it has good heat-transport qualities and does not cause fires, but it is expensive and prone to leakage. Carbon dioxide, an inexpensive substitute, is currently used in all gas-cooled reactors.

Gases have the disadvantage of low thermal conductivity and density. For these reasons, the flow rate must be high, and a lot of power is required just to move the coolant. Liquid coolants in a power reactor must flow at 15 to 20 feet per second (4.6 to 6.0 m/s). The gas coolant must flow 10 times faster.

Although the gas-cooled reactor is considered a historical relic, there are still seven units operational in the United Kingdom, and a few are running elsewhere in the world. As is the case with all reactors that can use natural uranium fuel, gas-cooled reactors can be refueled while in operation. A water-cooled reactor must be shut down and partially dismantled for refueling. This is a difficult task, requiring days or weeks of unproductive downtime. A gas-cooled reactor or a CANDU is loaded with new fuel by a machine on one side, while spent fuel falls out the other

UNUSUAL REACTOR COOLING STRATEGIES

There are many reasons why liquid sodium or a sodium-potassium alloy are unattractive reactor coolants, but there is one very advantageous aspect of a metallic fluid. A source of trouble in liquid cooling systems is the pump required to move the coolant through the reactor core. Even though a pump can be simplified down to one moving part, the impeller, it is still a problem. The impeller is a rotating component, designed to force the coolant to flow through the pipes like a windmill in reverse. It must be sealed, so that coolant will not leak past the rotating pump shaft, and the seal is a point of eventual failure. Failed seals and erosion on the impeller are potential problems in a cooling system.

Metallic coolant is unique in that it is an excellent electrical conductor. The coolant pump in a liquid metal system exploits this property. A conductor is forced to move if electricity is applied to it with a constant magnetic field running perpendicular to the direction of motion. There is thus no need to have an electric motor armature turn a shaft that turns an impeller. The coolant itself is the armature. The pump is simply a modified section of coolant pipe, with a current of electricity applied across a diameter. A steady magnetic field outside the pipe runs from top to bottom, and the liquid metal flows quickly. With no moving parts, this is an ideal coolant pump.

Experiments have been conducted to find alternatives to water as a coolant. A possibility was Dowtherm A, a mixture of biphenyl and diphenyl oxide, commonly used in the chemical industry as a high-temperature heat-transfer medium. The

side. The entire operation takes place while the reactor is running and generating power.

LOOPED COOLING SYSTEMS

The earliest power reactors, built during World War II to produce plutonium-239, were water cooled but used an open loop system. Water was drawn out of the Columbia River in Washington State, pumped through the reactor, and after cooling off in a retention pond, dumped back into the river downstream of the take-out point. This simple cooling strategy has been superseded by a closed loop system for multiple reasons.

organic moderated reactor experiment was constructed about 1957 to test this concept.

An advantage to this coolant was an automatic power-stability characteristic. It had a negative temperature coefficient, meaning that its moderating characteristics would deteriorate as the temperature climbed. This would tend to shut down the reactor if it was running too hot or power it up if it was running too cold, and this inherent control feature was good. Unfortunately, the organic substances that had proven so well behaved in chemical plants would break down quickly in the high-radiation environment of a nuclear reactor. The liquid coolant would solidify. In a reactor running at a power level of 240 megawatts, 260 pounds (118 kg) of coolant would decompose per hour. In a water cooled reactor, coolant does break down, but it does not become a worrisome solid that can block passages and gum up a valve.

A reactor running at a very high temperature, more than 1,200°F (649°C), produces more energy in less space than a reactor running at boiling water temperature. High-temperature coolants requiring no high-pressure containment have been tried. Molten fluoride salt experiments have proven that this exotic coolant can cool a very hot reactor. A mixture of lithium-fluoride and beryllium-fluoride has been used successfully. A special material, INOR-8, an alloy of nickel, molybdenum, chromium, and iron, is a necessary material for making pipes, valves, and pumps for this coolant. A molten salt reactor is feasible, but it suffers from similar disadvantages that plague the liquid metal coolants, and the viscous salt is difficult to move with a pump. There are no molten-salt power reactors currently in use.

The water coolant became contaminated with radioactive material as it flowed through the graphite reactor, which was then deposited in the river, causing a dangerous situation downstream. Coolant pipes would become activated by the strong neutron presence in the reactor, and pipes naturally erode away on the inside as water courses through. A break in a pipe could be disastrous, as highly radioactive fission products could directly contaminate the coolant.

The high power output of the plant was completely wasted, as all the heated water was introduced back into the river with no electrical power generation. The purpose of these early plants was to make plutonium, and the parallel production of power was discarded. This type of plant was uniquely located by a large river with enough flow rate to absorb the power produced by several reactors running at once, and this was a situation that could not be repeated everywhere.

Commercial power plants are now built to make hot water and turn it into electricity. Plutonium conversion in the fuel is usually discarded. Nothing radioactive is dumped into the environment on a continuous basis and, while running water is advantageous, smaller streams can be used, with a minimum of heating and no isotope residue. This revised configuration is possible by closing the water cooling loop. The same water is pumped around and around in the reactor system, alternately cooling the reactor, then driving the power turbines, and being cooled for reintroduction into the reactor.

The simplest implementation of the closed loop system in a nuclear power station is the boiling water reactor (BWR) configuration. In the BWR vessel, the uranium fuel is arranged in a neat lattice, surrounded by moderating water. The water is heated to boiling by the fission process, and the steam rises to the top of the vessel. Hot steam exits through a pipe near the top of the reactor and is conducted from the reactor containment structure, through an opening in the concrete wall, to the turbine deck.

The turbine is one long driveshaft, and on the same shaft are several stages of steam turbine wheels plus the electrical generator. The steam acts on the wheels to turn the shaft, and the generator armature turns with it, generating electricity.

The steam's energy is expended turning the turbine wheels, and the pressure drops. The steam exits the turbine housing through the floor of the turbine deck to the condenser underneath. The spent steam is cooled, and the last vestige of energy is removed. The steam condenses back into water, and the warm fluid is then conducted back into the reactor containment structure. A small quantity of loose hydrogen and oxygen are recombined back into water in a catalytic converter, and the water is forced back into the high-pressure environment of the reactor vessel by a large electrically driven pump.

The combination coolant and moderator, slightly contaminated with bits of fuel and *tritium,* never sees the outside of its closed environment. It circulates continuously, alternately being heated and cooled. There is one connection outside the plant, but it is indirect. All steam-driven electrical power plants, be they nuclear, coal, geothermal, or gas-driven, must have an ultimate heat sink.

THE ULTIMATE HEAT SINK

In each water-cooled nuclear plant, there are at least two coolant loops, connected in series. The first loop is entirely inside the plant, and it cycles water through the reactor, the turbine, and the condenser. The condenser acts similarly to an automobile radiator, which exposes the hot water to as much surface area of metal as is possible by distributing the flow among many smaller pipe flows. The heat is conducted from the fluid running in the small pipes to another medium outside the pipes. In the case of a power plant, the hot water inside the condenser pipes is cooled by water outside the pipes, transferring the heat from the inner loop to another water loop.

The iconic cooling towers at a nuclear power generating station. Although they are the largest structures on the site, they are mainly hollow tubes, with air and water vapor being blown upward through them. *(kated/Shutterstock)*

This water loop is semi-open, and another set of pumps moves this secondary coolant from the steam condenser to outside the plant to the cooling towers. The water is sprayed into the tower, where it is cooled by evaporation. Most of the cooled water is pumped back to the condenser, and water that is lost in evaporation is made up from the water source near the plant.

Cooling towers have become an icon of nuclear power, even though any type of steam plant uses exactly the same ultimate heat sink. A photo of a nuclear plant usually shows two tall structures having a characteristic hourglass shape, cylindrical but pinched in the middle. These are the cooling towers, designed to exhaust the excess heat that is always generated in power production into the greater environment.

In the simplest cooling tower design, the water from the steam condenser is spread into a horizontal grid of sprinklers, pointed down, in the middle of the tower. There is a natural draught chimney effect in the tower, causing air to be sucked in at the bottom and flow upward, out the top. The water sprinkles down through the moving air, and about 1.5 percent of it evaporates, making a visible plume of vapor flow out the top with the moving air. It looks as if the plant were manufacturing clouds. The cool water collects in a pool below the tower and is pumped from here back to the condenser in the plant. The water lost by evaporation is made up from water pumped from the nearby source, such as a river or ocean.

In larger, more complex secondary-loop cooling systems, the water from the steam condenser is circulated through a helix of pipe inside the tower, where it exchanges heat with the air. Cool air enters the bottom of the tower, which is elevated slightly off the ground. It travels up the tower, absorbing heat from the condenser, and it exits out the top, rising up into the atmosphere. A large electric fan at the bottom of the tower encourages the upward flow. The cooling of the condenser water must be augmented, and for this a source of flowing water is necessary. This last, series-connected cooling stage is always an open-loop configuration.

Water from the nearby source is pumped into the cooling tower. It is directed into a circular sprinkler system, where it is rained onto the heat-exchanger pipe in the tower. It evaporates upon hitting the pipe, further cooling the condenser water as it evaporates into the air. Water dripping off the pipe is directed back into the river, downstream of the take-out point. It is so far removed from the radioactive action in the reactor that it has not the slightest contamination.

EMERGENCY CORE COOLING SYSTEMS

The fact that there is a small component of nuclear fission that is delayed is good for easing the control of a reactor, but it adds a problem to the cooling system. One percent of a reactor's power production is on a delayed schedule, dropping off exponentially with time. Most of the delayed action is gone within 40 seconds, but there is a small component that goes on for days and even years. One percent of 1 billion watts, the typical power output of a nuclear plant, is 10 million watts, and this is still a lot of power. Entire power plants have been built that produce less power than the delayed power that is available after shutdown in a nuclear plant. This residual power, if not properly handled, is enough to melt the reactor core.

Any trouble in a water-moderated nuclear reactor will shut it down completely and quickly. Both automatic emergency shutdown systems and the basic nature of the reactor design will prevent it from generating power if there is a problem. Even in total shutdown mode, a power reactor still generates that residual power. This is managed by the normal reactor cooling system, which is kept on and running after a reactor is shut down.

However, anything can go wrong, including the loss of the primary cooling system. If the primary coolant pump and the backup coolant pump should fail, for example, then the coolant will stop running and the core can overheat. A nuclear power reactor is built with redundant backup systems in every aspect of the plant, and this includes the cooling system. This supplementary cooling is provided by the emergency core cooling system (ECCS).

Multiple subsystems make up the ECCS. The high-pressure coolant injection system is a set of electrically driven pumps that can spray water into the reactor vessel when the coolant level has dropped to a predetermined threshold. It is possible that a nuclear plant could lose electrical generating capacity and be shut out of the larger power network. In this case, these high-pressure pumps are kept operating by auxiliary diesel generators. These generators are always poised to take over the job of providing electricity to all the electrically driven equipment in the plant, including coolant pumps.

The depressurization system is a series of valves that can open automatically to vent steam off the pressurized reactor vessel. At this point in an emergency, the low-pressure coolant injection system can take over, adding coolant to the reactor vessel to keep the fuel from melting. A core-spray system in some reactors can inject water directly onto the fuel rods. This system can operate in either high-pressure or low-pressure situations.

The reactor containment structure, which surrounds the reactor vessel, can be cooled by the containment spray system. This is used to condense steam that has been vented out of the reactor vessel. In some reactor plant designs, there is an isolation cooling system that can work if the diesel generators have malfunctioned. This system consists of pumps that run directly on residual steam from the shutdown reactor.

The process of removing heat from a nuclear power reactor is thus controlled and managed during all situations, including dire emergencies in which multiple systems have failed catastrophically. All licensed reactors have this level of emergency backups, regardless of the type. There are a few standard types of reactor in use in the world today, all achieving the same result in remarkably different ways.

5 The Many Ways to Build a Power Plant

Several schemes have at least been tried in designing and building prototype nuclear power plants. One of the first plans was the Daniels Pile, a civilian power reactor at the Oak Ridge National Laboratory in Tennessee, named for its chief designer, Dr. Farrington Daniels (1889–1972). Although the reactor was never built, it was an elegant design named the pebble bed reactor, and it still holds promise as one of the most advanced reactor concepts being developed.

At the same time, immediately following World War II, the Soviet Union was developing the AM-1, or "Peaceful Atom," nuclear power plant in Obninsk, Russia. Although it produced only five megawatts of electricity, it may be considered the first nuclear power plant, producing electricity on June 26, 1954. The AM-1 was an odd combination of graphite moderator and boiling water coolant, and the design would mature into the Soviet RBMK reactors, a few of which are still operating.

There were other firsts. The first 400 watts of electricity produced by a nuclear process in 1951 was from the EBR-1 in Idaho, a fast breeder reactor experiment. The first significant electrical power, 200 megawatts, to be produced by nuclear means was generated by the Calder Hall reactors at Sellafield, England. The first Calder Hall reactor, using graphite moderator and gas cooling, was officially connected to the United Kingdom power grid on October 17, 1956.

Although all of these successful first reactor designs are now obsolete, the Daniels Pile, which was never built, inspired a naval officer named Hyman Rickover (1900–86) to plan the construction of a nuclear submarine reactor. This compact nuclear power unit was eventually built and installed in the world's first nuclear submarine, the USS *Nautilus,* in 1954. Although built under military secrecy for a very specific mission, this special reactor plan, the pressurized water reactor, would eventually rise to become the dominant design strategy of nuclear power plants all over the world.

THE PRESSURIZED WATER REACTOR

The reason for building the first pressurized water reactor (PWR) was based entirely on the narrow hull of a standard fleet submarine, which was only 28 feet (8.5 m) in diameter. There was no room for a graphite reactor, which would be as big as a house, or even a boiling water reactor, which would be twice as tall as a submarine. The size-dependent component of a nuclear power generating system was the reactor vessel, and a PWR vessel was comparatively small.

The reactor core was about the size of a garbage can, and it fit closely in a thick steel vessel, built to withstand the pressure needed to keep water from boiling away at high temperature. The fuel in a PWR can operate at temperatures up to 4,800°F (2,650°C) with the surrounding water at 550°F (288°C). Under normal atmospheric conditions, water boils away at 212°F (100°C), but the PWR vessel acts as a pressure cooker. It maintains the water in liquid state even though it is heated far beyond the boiling point. The superheated water exerts a pressure of more than 2,000 pounds per square inch (14,000 kp) on the inside of the reactor vessel, which must be constructed to withstand this stress.

Steam bubbles are voids that do not contribute to the neutron-moderating process. Having no steam bubbles in the water coolant means that the fuel is moderated by the water to the highest efficiency, so the uranium core can be kept to a minimum size. The fuel in a civilian PWR power plant is enriched to between 3 and 5 percent uranium-235. In a military submarine reactor, it is enriched to 50 percent. The higher percentage of uranium-235 maximizes the time between refueling shutdowns.

The fuel for a PWR in a civilian power plant is uranium oxide. Its high melting point of 5,100°F (2,800°C) makes it an appropriate material for the

Pressurized Water Reactor Vessel

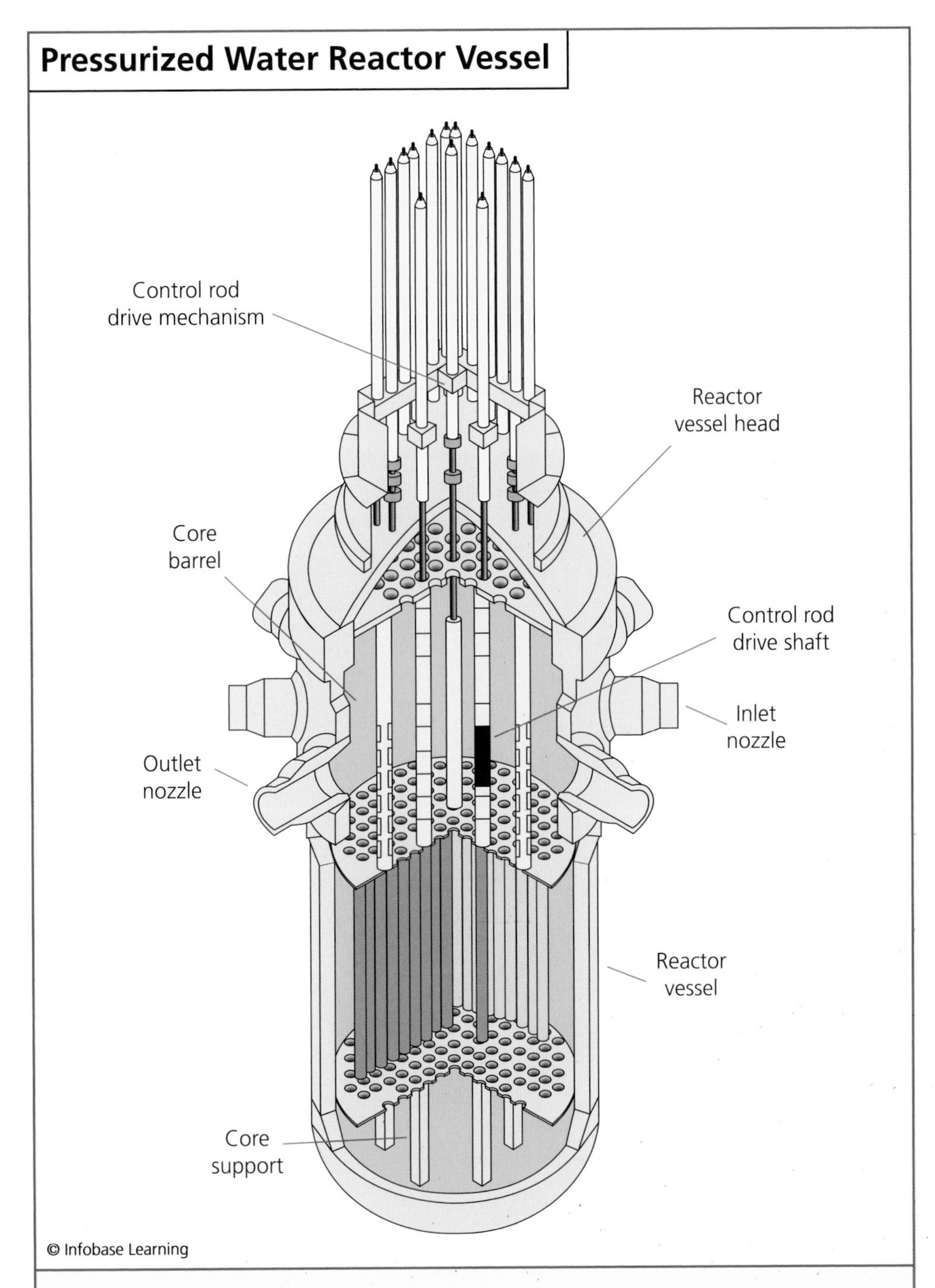

A cutaway diagram of the pressurized water reactor core vessel, showing the control rod drive mechanisms inserted through the top, into the uranium core. There are multiple penetrations of the vessel for water inlets and water outlets.

high-temperature conditions in a PWR, and it is not prone to combustion in air, as is the metallic form of uranium. PWR fuel is made by press-forming *enriched uranium* oxide powder into small pellets, about the diameter of a writing pen. Hundreds of these pellets are lined up and loaded into a metal tube, about 13 feet (4m) long. The tube is made of a strong high temperature–resistant zirconium alloy, zircaloy-4, with a very low neutron capture cross section.

The fuel rods are then grouped into square bundles of 200 to 300 rods each, with spacers made of zircaloy-4 keeping the rods secure. A large pressurized water power unit has 150 to 250 fuel assemblies

Pressurized water reactor vessel closure heads, ready for installation. Each hole in the rim is for a stud to hold it down on the vessel, with a nut tightened down on each stud. ***(NRC File Photo)***

loaded vertically into the reactor vessel, and this amounts to 88 to 110 tons (80 to 100 mt) of uranium. With that much uranium, a power reactor can produce 1 billion watts of power continuously for years.

For a PWR power plant, the refueling is on an 18-to-24-month cycle. In a refueling operation, one-third of the fuel bundles are removed, usually from the center of the reactor core. Peripheral bundles are moved to the center, and the new fuel is placed in the empty locations. It takes a few days to accomplish a refueling, even with a well-trained crew operating at high efficiency. To refuel a power reactor, the operation must be shut down completely, and the top of the reactor, or the closure head, must be carefully removed using a crane. The closure head is held to the top of the reactor vessel with typically 60 threaded nuts, all of which must be spun off with a large, remotely controlled socket wrench before it can be lifted.

Atop the closure head are located the control rod drives. There are typically 20 to 64 control rods in a PWR core. These electrically driven rods are actually long, four-lobed paddles. They are designed to fit in the spaces separating fuel bundles, soaking up neutrons and keeping the reactor core at a precise level of criticality. The controls are individually moved up and down in the reactor core under remote direction from the control room. Moving a control rod down into the core reduces the neutron activity, and moving it up and out of the core increases the neutron activity. As the uranium fuel is exhausted over the course of several years, the bank of control rods is gradually moved outward to compensate for the loss of fission activity from the fuel. To shut the reactor down, the controls are inserted fully into the core.

Control rods are made of a material having a very large neutron-absorption cross section. There are many exotic materials used for controls, from dysprosium titanate to zirconium diboride. The most common PWR control material is a silver-indium-cadmium alloy, encased in a stainless steel shell that prevents corrosion in hot water.

The metal control rods are used as a way to achieve complete or emergency shutdown and to selectively control the density of fission neutrons in the reactor core. However, most of the neutron control in a PWR comes not from the control rods but from boric acid, a neutron absorber dissolved in the water in the primary cooling loop. A secondary control system consists of the charging and letdown equipment. This system removes water from the high-pressure primary loop,

How a Power Plant Operates

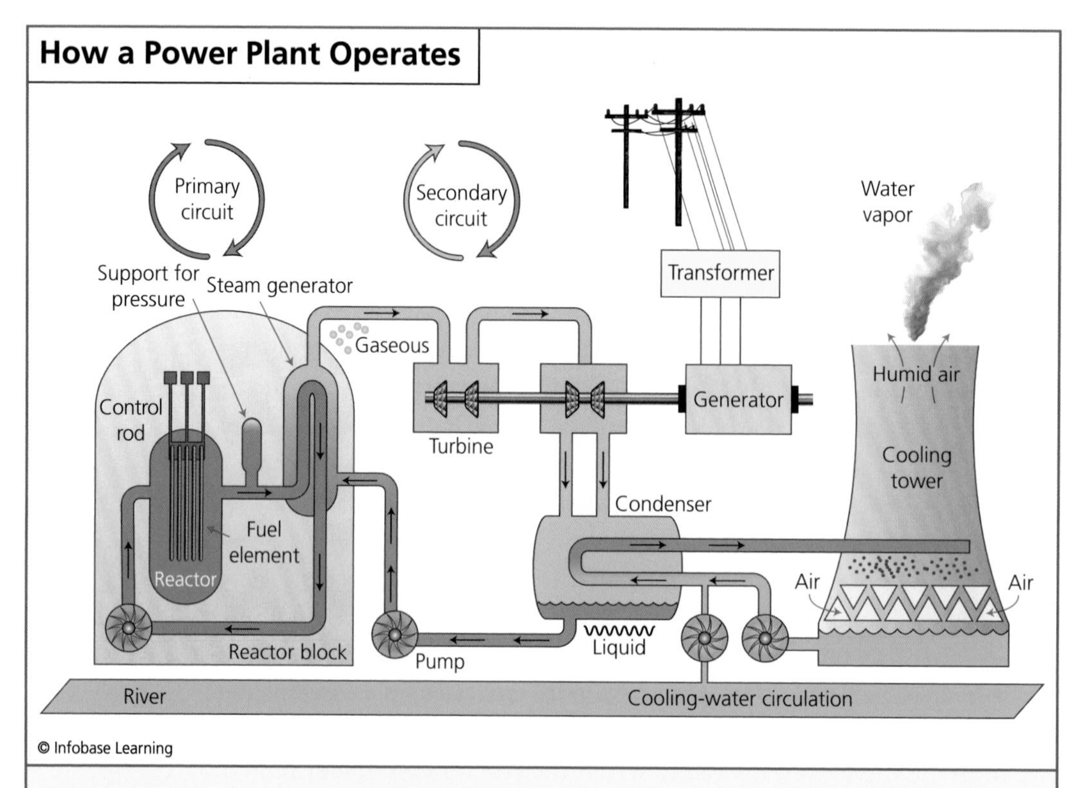

The PWR power plant system, showing the closed primary and secondary coolant loops, both of which use water as the coolant. The final step in cooling this reactor is an open loop, using water from a river to operate the cooling tower. This water is pumped from the river and escapes as vapor from the top of the cooling tower.

adjusts the concentration of boric acid in the water, and reinjects it back into the loop using high-pressure pumps. The boric acid control strategy is characteristic only of pressurized water reactors.

The PWR design is also unique for its double cooling loop, in which the inner closed loop is hot water. The inner loop is kept in constant liquid phase by maintaining high pressure in the reactor core. Under most operating conditions, the pressure is kept high by heat from the fission in the reactor core. To insure that the pressure is maintained under all transient conditions, an electric water heater, or pressurizer, is built into the primary coolant loop. Its thermostat cycles the electric heater element to keep the water at 653°F (345°C); at this temperature the water in the closed loop is always liquid.

The high-pressure hot water is diverted into four boilers, or steam generators, that exchange heat from the inner loop to an outer closed loop. The boilers make steam at 900 pounds per square inch (6,200 kp) and 530°F (275°C), which is used to drive the turbine and turn the generator. Although this is a complicated and expensive way to build a power plant, with two closed loops in series, it isolates the turbine equipment from any chance of radioactive contamination, and it has proven to be safe and reliable over 60 years of operations and design improvements. The PWR is the world's most used reactor design. Most countries using nuclear power generation have at least one PWR plant in operation.

The Westinghouse Electric Company has improved this traditional PWR design and is now selling the AP1000, a Generation III+ reactor hav-

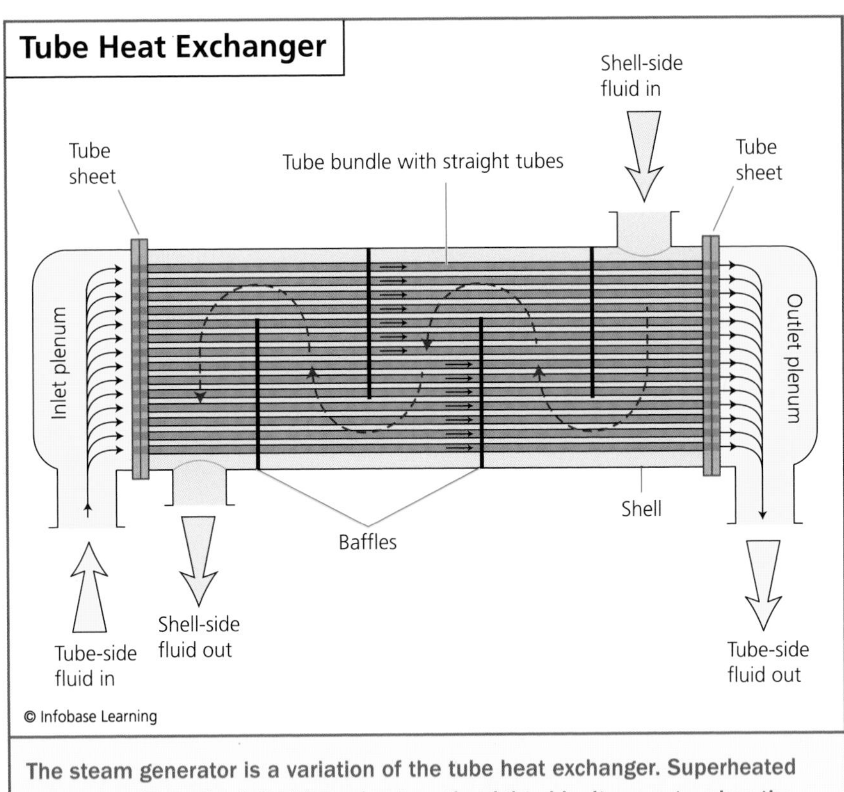

The steam generator is a variation of the tube heat exchanger. Superheated water goes in on the left side and out on the right side. It never touches the other circuit of water that comes in at the top and leaves as steam at the bottom.

ing several updates. The cooling system is still the double-loop of the submarine reactor, but safety backup systems, which provide additional coolant to the reactor core in cases of cooling-loop failures, have been made passive. These passive systems do not depend on electrical pumps or backup diesel electric generators. Gravity is used to put extra water where it is needed, and this simplification greatly improves the safety of the plant and reduces the potential cost of a mechanical breakdown. At this time four AP1000 reactors are being built in China, and several are planned for installation in the United States.

THE BOILING WATER REACTOR

The boiling water reactor (BWR) is distinguished from the PWR by the size of its reactor vessel and the fact that it has only one closed coolant loop. The commercial BWR was developed by the General Electric Corporation in the 1950s, and the first BWR installation was the Dresden 1 Nuclear Power Plant in Grundy County, Illinois, near the head of the Illinois River. It generated 210 megawatts of electricity starting in 1960 until final shutdown in 1978. There are now 93 BWR power plants operating in the world, with 35 in the United States. In addition to General Electric in the United States, BWRs have been built by Hitachi and Toshiba in Japan, Kraftwerk Union in Germany, and ASEA in Sweden.

The reactor vessel in a BWR plant must be large because water is boiled in the reactor core. The steam bubbles reduce the ability of the water to moderate neutrons, so the core must be larger to achieve criticality. However, this increase over the size of a PWR core does not mean that more fuel burn-up is required for power generation. Both the BWR and the PWR have about the same efficiency of uranium fissions per megawatt-hour of energy produced. Also, the BWR vessel must include a tall steam separator and steam dryer assembly over the core to keep liquid droplets out of the steam supply. The uranium core in a BWR is about 12.5 feet (3.8 m) high. The entire vessel, including the water droplet exclusion equipment at the top, is about 72 feet (22 m) tall and 21 feet (6.4 m) in diameter. These outer dimensions kept the BWR from being a submarine power plant and led to the development of the PWR.

The vessel is made of stainless steel–clad carbon steel, six inches (15 cm) thick. A one-piece closure head is bolted to the top, similar to the PWR design. The steam separators, located on top of the uranium core, are several hundred metal tubes. As steam rises in a tube it is made to spin by fixed turning vanes at the bottom. Centrifugal force in the spinning steam column causes the heavy water drops to hit the inner wall of the tube, and the water drips back down into the core. The dryer units are mounted on top of the steam separators, and they remove the last vestige of moisture from the steam by causing condensation on vertically mounted metal vanes. Water collects in a tray underneath the vanes and is dropped back through the steam separators and into the core. The dry steam exits the vessel through a single pipe outlet on the side and goes directly to the turbine.

A PWR steam generator with the covers off, showing the complicated tubing arrangement inside ***(NRC File Photo)***

A BWR fuel assembly consists of 74 to 100 fuel rods, very similar to the fuel assemblies used in PWRs and using basically the same fuel pellets. There are approximately 800 fuel assemblies in a BWR core, composed of 154 tons (140 mt) of uranium.

The BWR is controlled by two methods, metallic control rods and variable steam bubbles. The control rods are similar to those used in the PWR, but they must be inserted into the core from the bottom and not through the top of the reactor vessel, because the steam drying equipment crowds the top section. The active ingredient in the control rods is usually boron carbide, and they are moved in and out of the core by pneumatic or hydraulic pressure. Under emergency conditions, the entire bank of control rods can be inserted completely into the core in under four seconds. As in the case of the PWR, the metallic controls are used mainly to shape the flux profile in the reactor core, making the fuel burn up evenly to the edges of the core, with the fissions not concentrated in the center.

The more subtle control is through varying the speed of water as it flows through the core. Slowly moving water tends to pick up steam bubbles, and these bubbles are voids in the water moderator. Voids produce no moderation, and bubble-laden water tends to shut down the reactor. Fast-moving

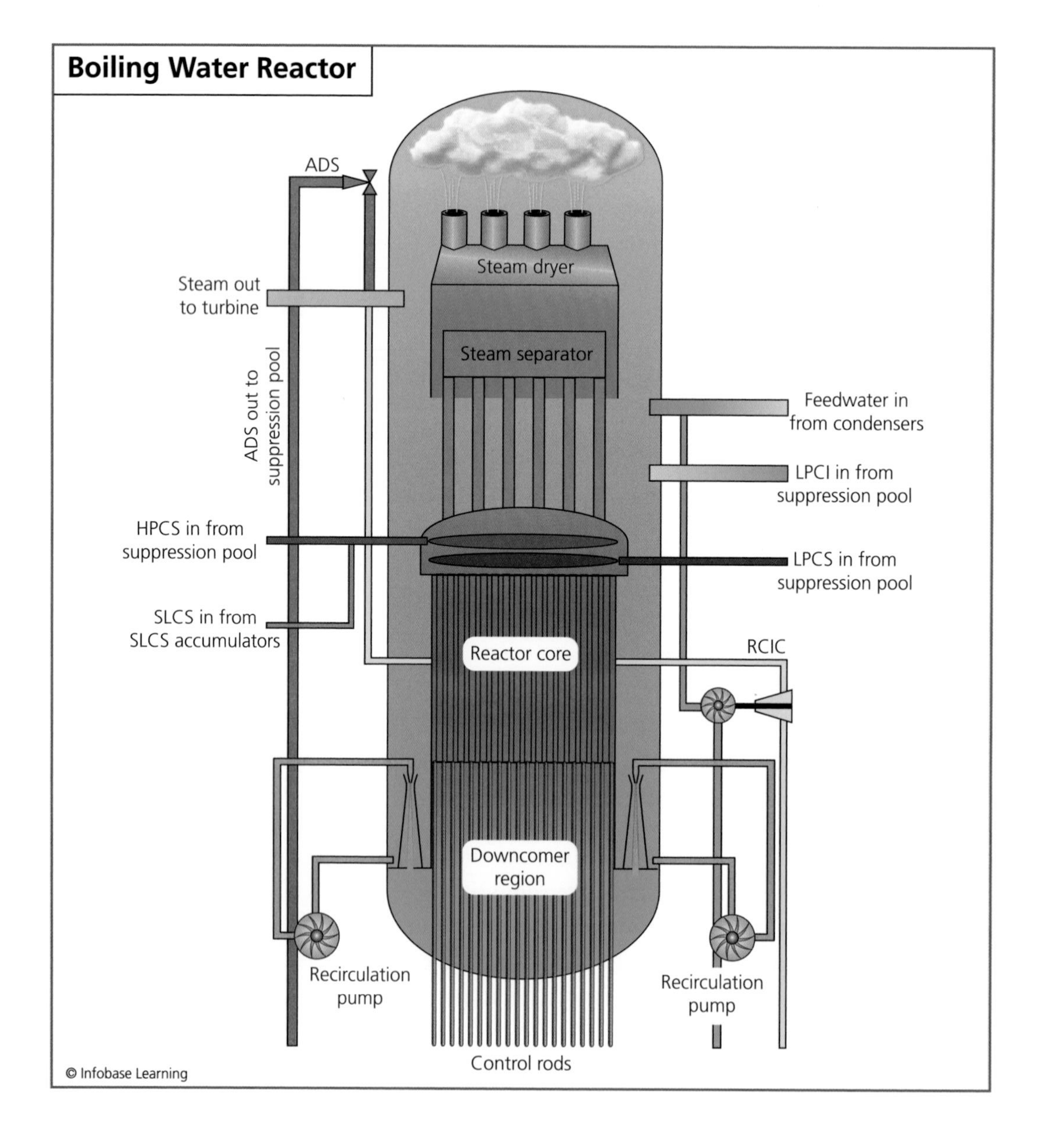

water, free of bubbles, is an efficient moderator, and the power in the core picks up under this condition. Most large General Electric BWRs have jet pumps in the reactor vessel, driven by variable-speed electrical pumps. Water is vented out the side of the vessel, driven to speed by the pumps, then reintroduced to the vessel above the core and blown through jet nozzles. By controlling the speed of these pumps, the bubble content of the boiling water is adjusted to the desired state of criticality in the reactor.

This jet pump control system has been improved and simplified by an alliance of General Electric and Hitachi Ltd., a high-technology services company in Tokyo, Japan. This new design is the advanced boiling water reactor (ABWR). The most notable feature of this Generation III reactor is the redesigned jet pumps. Instead of having complex piping outside the reactor vessel to recirculate the water through the core, there are 10 pumps located inside the vessel, with only the drive motors located outside. The traditional analog controls for the power plant have been replaced with digital controls and instruments. The plant is now controlled and monitored using computer terminals and has an autopilot option, in which the reactor can be automatically started, controlled at the power production level, and shut down without operator intervention. An ABWR power plant is designed for a 60-year life, and one reactor produces 1.35 billion watts of electricity.

There are presently four ABWRs in operation in Japan, and one is under construction. There are two ABWR plants in Taiwan.

General Electric has designed an even more advanced reactor, a Generation III+ plan named the economic simplified boiling water reactor (ESBWR). Although there have been no ESBWRs built, it is a bold design, unique in its lack of any coolant recirculation pumps. The cooling system is gravity driven, relying on the fact that hot water is lighter than cold water to maintain the circulation. This design is attractive because circulating pumps are expensive, complicated, and prone to failure, all of which are eliminated by the natural circulation characteristics of the ESBWR cooling system.

The first NGNP reactor may be operational as early as 2021.

(*Opposite page*) ***A schematic representation of the BWR vessel.*** **The ADS (automatic depressurization system), HPCS (high pressure coolant spray), SLCS (standby liquid control system), LPCI (low pressure core spray), and RCIC (reactor core isolation cooling) are all components of the complex system designed to protect the core from melting in any foreseen condition of emergency or complete breakdown.**

THE GAS-COOLED REACTOR

The 60-year reign of the gas-cooled reactor (GCR) came to a close in 2010. The last examples of these power reactors, which are considered to be Generation I designs, are being closed down and dismantled. Although the GCR concept is technically obsolete, there are certain advantages to gas cooling, and the design may be upgraded in the future as part of the Generation IV power reactors in the next 20 years.

Although there are different GCR designs, all have in common a solid graphite moderator, natural or unenriched uranium fuel, and gaseous carbon dioxide coolant. The natural uranium fuel was important in European countries in the early days of nuclear power, as uranium-enriching facilities

REACTOR DESIGNS THAT ARE NOT USED FOR COMMERCIAL POWER

A set of theoretical power reactor designs is currently being researched by a forum of 10 countries with nuclear power generation, including the United States. These Generation IV reactors are not expected to be available for commercial power generation before 2030, with the possible exception of the very high temperature reactor (VHTR), also called the next generation nuclear plant (NGNP).

The current design of the NGNP is a variation of one of the world's oldest nuclear reactor concepts, the pebble bed reactor, originally designed at the Oak Ridge National Laboratory in Tennessee immediately after World War II. In the pebble bed reactor, the uranium fuel is formed into tiny spheres called pebbles. Each sphere is coated with a hard ceramic layer of silicon carbide. The carbon in the coating acts as the moderator, and the heat is removed from the fuel by blowing a gas such as helium through the mass of pebbles. A single fuel loading consists of 360,000 pebbles, and each pebble is about 2.6 inches (60 mm) in diameter.

The very high temperature of 1,832°F (1,000°C) that is possible with a pebble bed configuration makes it possible to create hydrogen from water with a thermochemical process. A fixed, stationary, pollution-free nuclear plant could therefore produce hydrogen to be used in pollution-free automobiles, and this is an attractive option for future transportation strategies.

A potential problem with the pebble-bed concept is one that has plagued graphite-moderated reactors in the past. If the sealed helium coolant cycle is breached, then were used only in the United States and the Soviet Union, and the goal was to be energy independent of these superpowers. The use of natural uranium led to dependence on purified graphite as the moderator. For the sake of safety and economy, carbon dioxide was chosen as the coolant. Unlike water, carbon dioxide does not build up explosively high steam pressure in the reactor vessel, nor does it absorb neutrons unproductively.

Carbon dioxide, heated to a high temperature in the reactor core, can be circulated repeatedly through the core in a closed loop. A second closed loop creates steam using heat exchangers, as is done in pressurized water reactors. Most GCR vessels are spherical and as large as a two-story house, due to the low power density in a natural uranium reactor having only

air can get to the hot graphite. Hot graphite catches fire easily, and it is hard to put the fire out.

China has the only working prototype pebble bed NGNP at this time, the HTR-10. It is currently being upgraded to a 200-megawatt electrical generating plant in Shandon Province.

Another advanced concept on the table at the Generation IV forum is the molten salt reactor (MSR). This is again a very old design that is being reconsidered. It was originally part of the aircraft nuclear power program and again it was designed at Oak Ridge, Tennessee, in 1954. The MSR is unique in that the fuel is dissolved in the coolant, which happens to be molten fluoride salt. In the Oak Ridge MSR, the fuel was uranium tetrafluoride, or green salt. This exotic mixture barely melts at high temperature and produces no high-pressure situation as does steam. The moderator is a solid graphite block, with holes drilled in it to allow the coolant-fuel mixture to flow through.

As fuel, both of these advanced reactors can use thorium-232, which is hundreds of times more abundant than uranium. Both suffer from the possibility of graphite catching fire, but these and other Generation IV reactor concepts have the following three overriding advantages over current power reactor designs:

- The fission product from these reactors will be radioactive for decades, instead of millennia.
- Efficiency will be greater, giving 100 to 300 times more electrical energy for the same amount of nuclear fuel.
- Existing nuclear waste may be consumed to produce electricity in these plants.

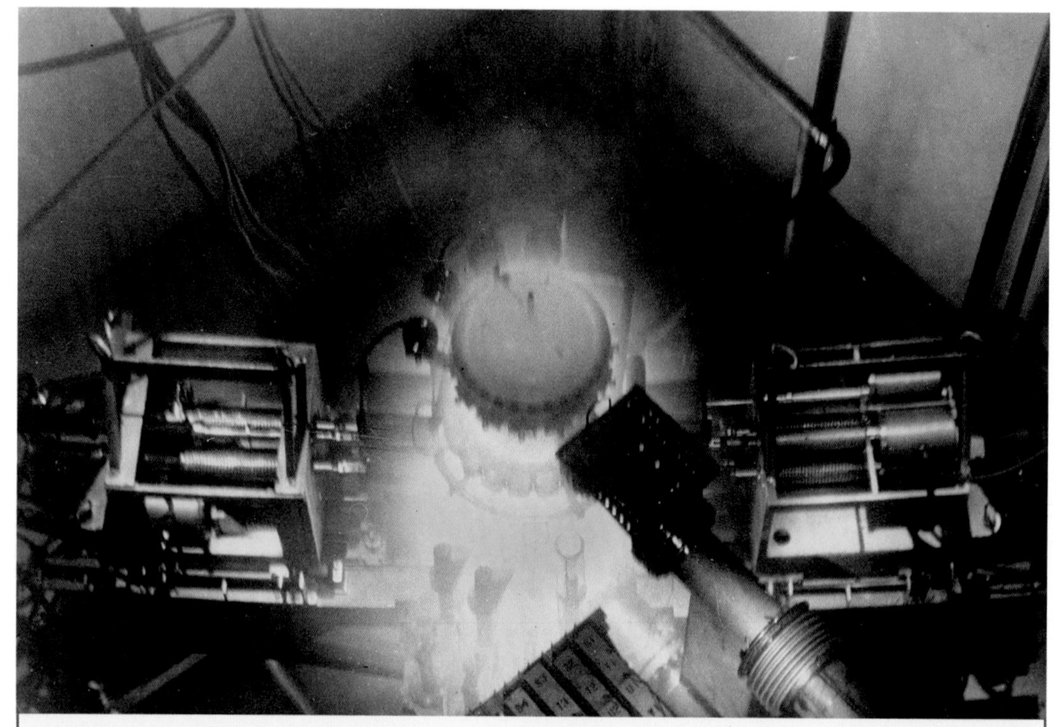

An early gas-cooled reactor experiment, running at its full power of two megawatts at the National Reactor Testing Station, Idaho, in 1962 *(DOE Photo)*

0.7 percent fissionable uranium-235 in the fuel. Control rods are inserted at the top of the reactor core and made of a boron-steel alloy. The reactor vessels can be made of steel or reinforced concrete, designed to handle gas pressure up to 392 pounds per square inch (2,500 kp).

The most successful GCR design was the *MAGNOX,* with 11 power stations built in the United Kingdom. A MAGNOX reactor was also exported to Tokai Mura in Japan, and another to Latina in Italy. North Korea took the declassified plans for a MAGNOX and built a copy. The first MAGNOX reactor, the Calder Hall Power Station at Sellafield, England, was first connected to the power grid on August 27, 1956, and it generated power reliably until final shutdown on March 31, 2003. Calder Hall had supplied electrical power for nearly 47 years.

The name MAGNOX comes from the fuel configuration of the reactor. The fuel is natural uranium, and its fission probability is optimized by fabricating it as metallic uranium, and not uranium oxide. The uranium metal is flammable, so it must be sealed against oxygen. This is done by encasing each cylindrical fuel element in an airtight magnesium-aluminum alloy can. MAGNOX is an acronym meaning MAGnesium Non-OXidizing. The magnesium is used because it has a very low neutron absorption cross section.

There are two disadvantages to this fuel design. First, the magnesium begins to corrode in the carbon dioxide above 680°F (360°C), which limits the power level of the reactor to about 160 megawatts. Second, it also corrodes in water and that prevents long-term storage of spent fuel underwater. Most spent reactor fuel is stored in pools of water, as water is an excellent shielding material for highly radioactive fission products and keeps the fuel cool as it decays over the course of several years.

In parallel with the British development of the MAGNOX reactor, the uranium naturel graphite gaz (UNGG) reactor was developed in France in the 1950s. Although it is very similar to the MAGNOX design, it differs

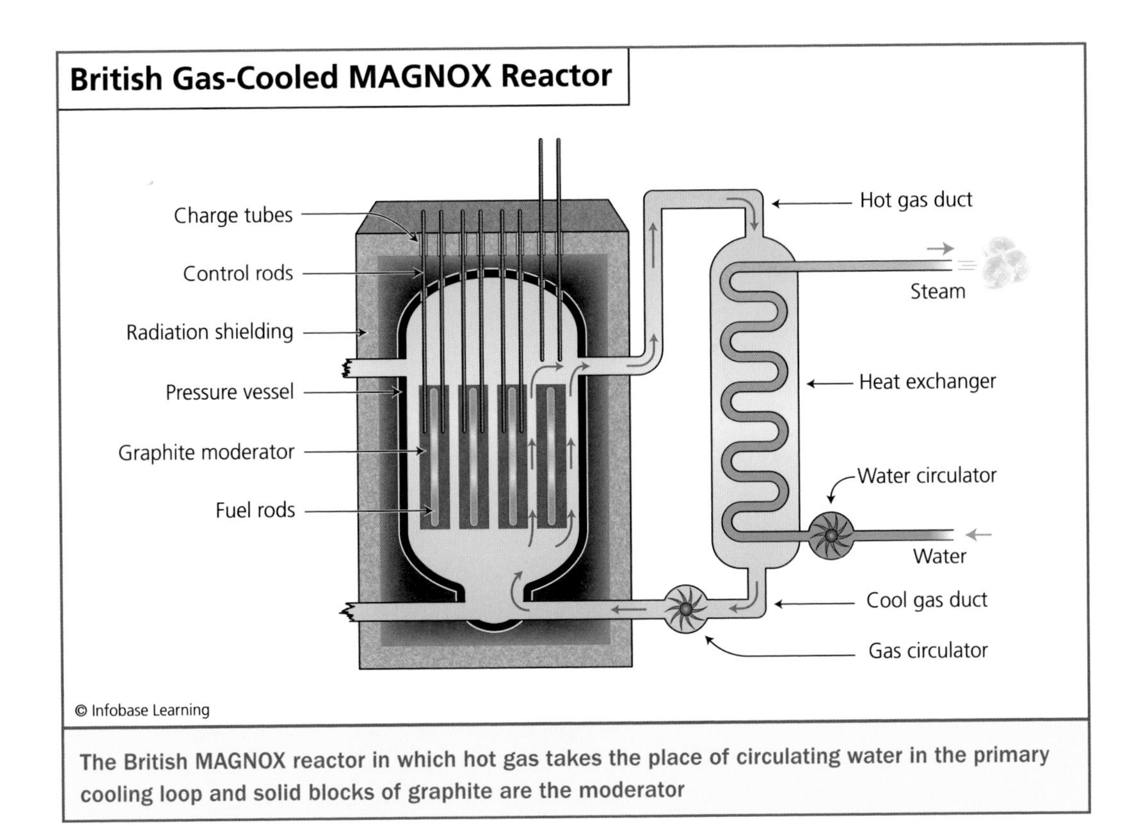

The British MAGNOX reactor in which hot gas takes the place of circulating water in the primary cooling loop and solid blocks of graphite are the moderator

A 1:50 scale model of the Heysham II advanced gas-cooled reactor in Lancashire, England, showing the reactor core in cross section. This reactor began commercial operation in 1989 and is scheduled for final shutdown in 2023. *(© SSPL/The Image Works)*

in two ways. The fuel in the French UNGG reactor is loaded horizontally, whereas the MAGNOX fuel is loaded vertically. The horizontal fuel tubes simplified the reactor refueling, which must be performed on a monthly basis in a gas-cooled reactor. The fuel cladding was a magnesium-zirconium alloy, which could operate at higher temperatures, but it also had corrosion problems in water. Of 10 UNGG units built and operated, all are now shut down and dismantled.

THE LIQUID METAL–COOLED REACTOR

Metal seems an odd fluid to circulate through a nuclear reactor core. It has an obvious disadvantage. When the reactor is shut down and has cooled off, the fuel is encased in solid, opaque metal. Not only is it hard to remove the fuel under this condition, but it is impossible to see through the metal for a simple inspection of the core. However, even with this shortcoming, metal has proven to have three advantages as a reactor coolant.

Metal liquefied under high temperature is an excellent heat conductor, and a reactor equipped with it can run at extremely high temperatures without generating a great deal of gas or steam pressure. A broken coolant pipe in a liquid metal reactor does not cause a destructive explosion, as does a broken steam pipe. The metal coolant provides negligible moderating effect. For a certain type of reactor design, this can be an advantage. Having to run a reactor on fast, unmoderated neutrons means that purified uranium-235 or straight plutonium-239 must be used as fuel, but it also means that a surrounding blanket of uranium-238 can be converted into plutonium-239 during power operation. At the high neutron speed of a metal-cooled reactor, the conversion is so efficient, more fuel is made than is burned. This desirable effect is used in a special reactor design, the liquid metal fast breeder reactor (LMFBR).

Five metals have been tried as reactor coolants. Each has its advantages and disadvantages. The first obvious choice was mercury, which is liquid at room temperature and can be easily drained from a cooled reactor. It was used in the first liquid metal reactor, Clementine, built in a cavern in a canyon in Los Alamos, New Mexico, in 1945. The reactor operators were called forty-niners as the military code word for plutonium was 49, and plutonium-239 was used as the fuel. Mercury has since fallen out of favor as a coolant because of its high toxicity and very high vapor pressure, even at room temperature. At high temperature, it produces noxious fumes and has a high neutron absorption probability.

Lead has been used as a reactor coolant. It has good nuclear properties and is a superb shield for *gamma rays,* which are produced in dangerous quantities during fission. However, lead has a high melting point of 621°F (328°C), and this exacerbates the refueling and inspection problems associated with metal coolants. A lead-bismuth mixture, or eutectic, has been used successfully as an alternative lead-based coolant. The melting point of this material is lowered to 254°F (124°C) and is considered in the Generation IV reactor design initiative as an advanced LMFBR coolant. A disadvantage is that the lead-bismuth corrodes steel, which is a major material used in power plant construction.

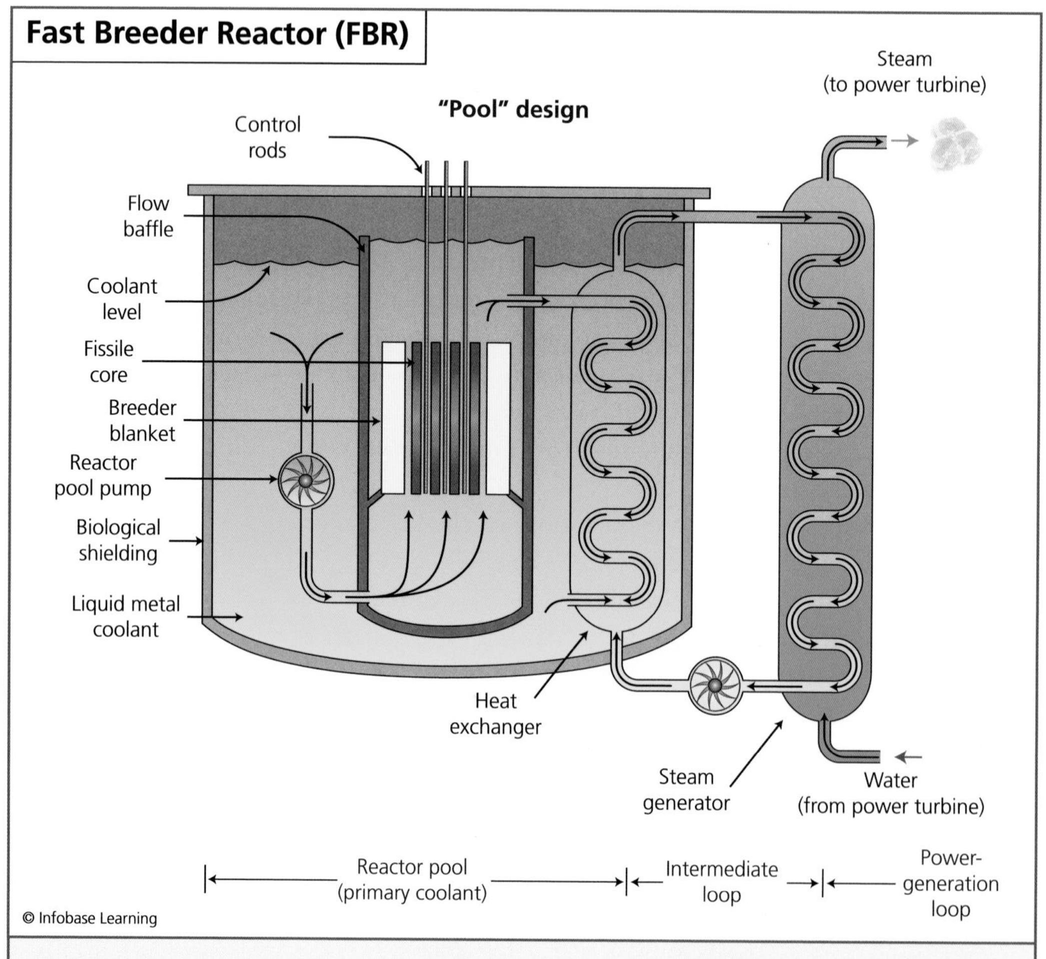

A schematic representation of the pool-type fast breeder reactor. This design uses three cooling loops, an inside loop using liquid metal, an intermediate loop using liquid metal, and a final loop for turning water into steam.

Sodium is the most often used coolant in liquid metal–cooled reactors. It does not corrode steel and has a melting point of only 208°F (98°C). It has been alloyed with potassium, which lowers its melting point considerably, to a very favorable 12°F (-11°C). This alloy, designated NaK, is thus liquid at room temperature, without the toxicity of mercury. However, sodium and NaK both ignite spontaneously in air and react explosively with water. The product of the reaction with water is sodium hydroxide, which aggressively corrodes aluminum.

Given these choices of liquid metal coolants, many experimental LMFBR power plants have been built. In the United States, the first commercial power to be generated by a liquid metal reactor was at the Santa Susana Field Laboratory near Moorpark, California. The sodium reactor experiment began operation in 1957, generating 6.6 megawatts. This reactor suffered from a problem that would plague experimental liquid metal reactors for the next few decades. A disturbance in the coolant flow caused a partial meltdown of the reactor core, shutting down the plant for an extensive rebuild. A similar accidental coolant-flow interruption caused a partial fuel meltdown in the Fermi 1 LMFBR power plant in Monroe County, Michigan, soon after it began power production in 1963. Plans for a commercial liquid metal power plant in the United States ended in 1983, with the cancellation of the Clinch River fast breeder reactor project near Oak Ridge, Tennessee.

In the far north end of Scotland, the United Kingdom *Atomic Energy Authority* operated the Dounreay fast reactor from 1959 to 1977. The Dounreay produced a modest 14 megawatts of power using NaK coolant and was replaced by the more powerful prototype fast reactor in 1975. This reactor produced 250 megawatts of electricity until final shutdown in 1994. Both experiments were considered successful, but there have been no firm plans for further liquid metal reactors in the United Kingdom.

France, Japan, Russia, and India have engaged in liquid metal–reactor development. France has built the largest such reactor, the Superphénix, a 1,200-megawatt power plant that entered service in 1984. This plant was shut down in 1997 due to a mixture of technical and political problems. A smaller, 233-megawatt reactor, the Phénix, has been in service since 1973 and, in addition to electrical power generation, it is being used to test nuclear waste deactivation by transmutation. By activating nuclear waste with excess neutrons in the liquid metal reactor, it may be possible to cause the fission products to decay quickly to harmless materials.

The Soviet Union built two full-scale liquid metal reactor power plants. The BN-600, a sodium-cooled reactor with triple cooling loops, has produced 560 megawatts of electrical power since startup in 1980. An inner sodium cooling loop is located entirely within the reactor vessel, and this transfers heat to a second sodium loop that penetrates the vessel at input and output points. This heated sodium then transfers heat to a steam generator, making steam for the power turbine.

Japan has built a demonstration liquid metal plant in Truruga, Fukui Prefecture, operating at 280 megawatts; it may be a triple-loop copy of the Soviet BN-600. Although it was first started in 1994, it was closed down in 1995 following a sodium leak in a cooling loop and is presently in standby condition.

India is engaged in the construction of a 500-megawatt prototype LMFBR at Kalpakkam; China is working on a 25-megawatt example; and South Korea is developing a standardized, modular liquid metal reactor. While this type of reactor is not yet part of the established world nuclear power source, it is definitely in the set of Generation IV reactors that may be built in the next 30 years.

As is the case with all reactor types, successful integration into the world power network depends on not only the reactor core, its cooling system, and its controls, but on every component in the power plant. These pieces are usually of standard proven design, taken from other long-established technologies. All of these parts, from the steam turbine to the power distribution equipment, work in harmony to produce electricity.

6 Nonnuclear Components of a Power Plant

With the exception of the fission reactor, the equipment necessary to produce electrical power in a nuclear plant was developed in the 19th century. It was ready, tested, and proven by the late 1940s, when the idea to apply nuclear heat to the power production process was first considered.

The nuclear reactor was and still is a technology in its infancy. To this day, no two power reactors are built the same. There have been attempts to arrive at a standard design, but nuclear technology is still being improved, and each new reactor is a step forward, using lessons learned from the last reactor to achieve better safety, efficiency, and production cost. It seems as if there is always a better reactor design in the next generation.

Problems of materials, construction techniques, and safe designs were ironed out of the rest of the plant before the first reactor was ever built. In the first half of the 20th century, burning coal and water reservoirs were the preferred power sources, and in decades of use the machinery for these power plants was perfected and tuned. Rather than complicate a new art with unproven equipment, as much was taken from existing power plant components as was possible for the fresh nuclear technology.

This chapter reveals the other components that make up an electrical power–generating station, apart from the nuclear reactor.

INSTRUMENTS AND CONTROLS

As is the case with most systems in a nuclear plant, the instruments and controls used to interface with human operators and even the concept of human operation were first tried and perfected in coal-fired power plants. Most nuclear plants in the world were built in the 1970s, and the control room layouts and controls were based on well-established but antiquated switches, levers, warning lights, and pen chart recorders that had been used for decades. By that time, coal plants had begun to upgrade to more compact, computer-driven display systems and automated controls, but by its nature nuclear power had to remain one generation back in the development process. Before it could be designed into a nuclear power plant, a system had to have been proven and perfected in a coal plant.

All operations in a nuclear power plant are controlled and monitored in a central location, a windowless room about 50 by 50 feet (15 by 15 m)

A section of a traditional control room panel in a nuclear power plant. Although future reactors will use more computer control and digital imaging, most nuclear plants currently use this style of manual control and process monitoring. *(NRC File Photo)*

in floor area called the control room. In some large plants with more than one reactor, everything is consolidated into one room, and the control room floor area can be as large as 50 by 100 feet (15 by 30 m). Each reactor and its power-generating equipment is run by a control room supervisor and two or three licensed reactor operators.

The control room layout consists of a combination of vertical panels and control benches, forming the walls around a central desk area. Less-used controls and indicators are on subpanels, located in back of the main control board and not in the immediate purview of the operator. At the very top of the vertical panels are the annunciators, which are square translucent plastic tiles arranged in rectangular matrices. Printed in the center of each tile is the title of an alarm condition, such as a temperature or a radiation level that is out of acceptable range. Alarm conditions pertaining to a specific area of the plant or type of machinery are grouped together in each rectangular arrangement. There can be hundreds of individual tiles.

Each tile has an electric light or multiple lights behind it. When the condition indicated by the tile's title is important to the operation of the plant, the tile is lighted up. When a situation of concern develops, the light behind the appropriate tile begins to flash, drawing the operator's attention to the condition. At the same time, a warning buzzer sounds, giving aural as well as visual indication of an operating parameter out of bounds. The operator can stop the flashing and the buzzing only by pressing an acknowledgment button. If the alarm condition still exists, the tile light turns red, and it will remain so until its monitored parameter is returned to an acceptable operating range. These annunciator tiles are a primary indicator to the operators of the status of all systems in the power plant.

On the vertical panels, at eye level, are the monitored parameters, or instruments showing the value of variable data. Data can be shown with a meter, similar to a speedometer, a digital number readout, or a simple indicator light such as the CHECK ENGINE light on an automobile. The following types of information are monitored:

- Pressure of fluids and gases, particularly steam pressure. Strict pressure limits are imposed on everything containing steam, particularly reactor vessels and heat exchangers, to ensure against destructive mechanical failures.
- Fluid flow through every system in the basic heat-transfer cycles and nearly all auxiliary systems. Maintaining a balance and

compensating for changes in power demand require a constant monitoring of fluid flows.

- Fluid level in the reactor vessel, steam generators, storage tanks, condensers, and waste sumps. The fluid level in the reactor vessel is perhaps the most important parameter to be monitored in a nuclear power plant. Having the coolant level fall below the top of the fuel in the reactor core causes irreversible damage to the plant, as the fuel melts.
- Neutron flux, or the power level in the nuclear reactor. The level of neutron activity is monitored at several locations within and outside the reactor vessel. Not only is the neutron level monitored, but the rate of change in the neutron level is also important.
- Temperature of the reactor vessel and coolant systems. Temperature, along with neutron flux, coolant flow rate, and coolant pressure, are all monitored as a protection against damage to the plant systems, particularly the core of the reactor and its fuel load.
- Component status, for hundreds of valves, pumps, and circuit breakers. These are usually simple indicator lights. A red light means that a valve is open; green that it is closed. If both lights are on, it means that the valve is in the process of being opened or closed. For a circuit breaker, or electrical switch, red means that it is closed, and green means that it is open. If a pump is turned on, its light is red, and if it is turned off, its light is green.
- Water chemistry in the coolant. The acidity and electrical conductivity of the water in all coolant loops are monitored to make certain that pipes, valves, and pumps are not being corroded from the inside.
- Electrical measurements. A power plant generates electricity, and it is important to monitor the voltage, the current, and the alternating phase of the generator, as well as the status of the off-site electrical distribution systems and the immediate electrical load on the system.
- Process and area radiation monitoring. The power plant operates on nuclear fission, which produces a great deal of radiation. Radiation is dangerous to the operating staff of the plant and to the general population if it should break free of the safety shielding, and it is thus monitored continuously all over the plant and for miles around.

Some of these system variables are continuously recorded, and in the traditional nuclear power plant control room pen chart recorders may still be used. A roll of paper slowly unreels, moves under a marking pen, and is respooled on another roll. The pen moves as would the needle on a meter, making a permanent record of the parameter being measured. Although considered archaic, this method does leave a permanent record of important parameters that cannot be accidentally erased or lost and has proven to be of considerable value for event reconstructions in accident analysis. Magnetic or digital media are considered a more advanced state of the art, but nothing can match the robust nature of a pen chart recording.

It is now possible to control a nuclear power plant entirely by autopilot. A computer system can start up a reactor, bring it to operating temperature, switch in the turbo generator, and maintain a desired level of power over varying load conditions, all without human interaction. Most power plants, however, are manually controlled, and all must at least have full manual backup. These controls are switches, push buttons, and dials, located on the horizontal benches against the walls of the control panels.

A switch actuator on the bench is typically a substantial shaft rising vertically out of the panel and bent at a right angle to form a handle. The handle is grasped in the hand and twisted clockwise or counterclockwise. If the operator lets go of the handle, a spring always returns it to the neutral condition. This type switch is commonly used to open or close a remote valve, and a counterclockwise twist of the handle always closes a valve. A clockwise twist always opens a valve. This operation is so standardized, there is no need for an operator to glance at the switch to know how to operate it for any valve in the plant. Component status lights are associated with each switch, both acknowledging the switch action and confirming completion of the requested action. Close a valve with a switch, and a light will show that the switch has been twisted counterclockwise. An additional light will show that the valve has actually closed, after a delay.

Another type of control is the push button. A common use of the push button control is to reset an automatic function. The safety systems in a nuclear plant will automatically protect the plant if a monitored parameter, such as pressure, temperature, or radiation level, exceeds a set point. If a set point overrun has tripped a safety system, it will automatically shut a valve or do something to prevent undesirable actions. Once the situation has been changed to prevent a set point overrun, the system must be reset with a push button. Indicator lights are associated with each button to acknowledge a reset and that normal running has been resumed.

Analog controls allow fine adjustment of some parameters, particularly coolant flows. Instead of having two conditions, opened or closed, some valves in the coolant systems as well as pump speeds are continuously variable. Although these flows are usually under automatic control, it is possible to override any system and control it by hand by turning a knob. A clockwise turn always increases flow and a counterclockwise turn decreases it. Operator feedback for analog con-

THE RETRO DESIGN OF A NUCLEAR PLANT

The electric power industry is rooted firmly in the past, and much of the equipment used to generate and transmit power to the electrical consumer was perfected many years ago. Most high-voltage power transmission is still done by wires strung over wooden towers, the same as it was 100 years ago. House current in the United States has been standardized at 110 volts, 60 hertz, since George Westinghouse (1846–1914) and Thomas Edison (1847–1931) waged the war of currents in the late 1880s to determine whether the country would run on alternating or direct current. Westinghouse's alternating current won the war. Transformers, cables, circuit breakers, and connectors were invented in a glorious flurry of American innovation, and much of this equipment was so well designed there has been no reason to change it.

The nuclear power plant is the latest in a series of incremental improvements of generating commercial electricity. The electrical wind turbine may seem a newer innovation and is presently causing much excitement as a renewable energy source, but the first large one was built in 1888 in Cleveland, Ohio. A megawatt-class, three-blade wind turbine supplied electricity to Vermont in 1941. The first privately owned commercial nuclear power plant was the Shippingport Atomic Power Station in Pennsylvania. It started supplying power in 1957. The plant was built with the latest nuclear technology, as designed for a Navy aircraft carrier, but beyond the reactor cooling system the plant was built with the traditional steam plant multistage turbine, as invented by Sir Charles Parsons (1854–1931) in 1884.

This seemingly retro design of a nuclear power plant is perhaps most apparent in the control room. Most nuclear plants in the United States were designed in the

trol action is provided by meters on the vertical board, monitoring all adjustable parameters and always using safety set points to prevent bad settings.

Controls and indicators are grouped by similarity, and the horseshoe-shaped collection of panels and benches is usually divided into the following three sections:

- The reactor system, including reactor coolant, reactor control systems, nuclear instrumentation to monitor reactor power, and the steam generators

1960s and built in the 1970s, and their control room equipment was of traditional design, using established practices and thoroughly proven components. Coal-fired steam power plants have been built for the past century and are still being built around the world. At about the time the nuclear plant constructions were being completed, there was a flurry of innovation in coal plant control room design. It seems ironic that the older coal technology has been significantly upgraded to computerized controls and instruments, while nuclear plants are still stuck in designs from the 1960s.

Nuclear plants in Japan and Europe are basically at the same level of sophistication, with some exceptions. Japanese nuclear plant control rooms are typically equipped with several television monitors, connected to television cameras in inaccessible parts of the plant, to monitor conditions visually and in real time. Russian control rooms make use of mimics, which are diagrammatic display panels showing schematic representations of piping systems. On the large colorful board, the light indicators are positioned over diagrams of actual pumps. At a glance, an operator can relate an alarm condition, indicated by a light, to an actual pump or valve in the system.

The traditional nuclear plant is on the verge of a complete makeover. New, modernized designs of the widely used PWR and BWR power plants are ready to be installed worldwide, and these plants will have computer controls. Keyboards, joysticks, and even mice will replace the switches, buttons, and dials, and color flat-screen displays will replace the annunciator panels and the pen charts. The new controls and indicators will be more compact, easier to use, and a great deal less expensive. Nuclear power will have caught up with coal plants, after a long term of hibernation.

- The reactor coolant support and emergency systems, including the emergency core cooling system
- Secondary cooling cycle systems, including the turbo generator, the steam condenser, and the external cooling tower system

This array of hundreds of dials, indicators, annunciators, and switches requires experience and operator training. An operator must know instinctively where every switch is located and the meaning of each annunciated alarm condition.

EXPLOSION CONTAINMENT STRATEGIES

There is no chance of a nuclear explosion in a nuclear power plant. A nuclear reactor and a nuclear weapon use neutrons in different ways to produce fission, and it is no more possible to have a reactor plant undergo thermonuclear explosion than it is to generate power with an atomic bomb. A nuclear weapon uses prompt neutrons at high speed in a hypercritical situation. A power reactor uses delayed neutrons at thermal speed in an exactly critical situation.

However, that is not to say that a nuclear plant is incapable of experiencing an explosion. Nuclear plants make steam as the means to turn a turbine generator combination, and steam explosions have been wrecking power plants, boats, and factories since the discovery of steam power in the 18th century. Steam is notorious for building up an enormous amount of pent-up energy that can release all at once in the occasion of an equipment failure, sending roofs, flywheels, and entire buildings into the air and flattening everything within a large radius. An invisible weak spot in a pipe, a boiler, or a valve can suddenly let go. In the case of a nuclear plant, such an accident has an additional danger. If a steam explosion should involve the core of the reactor, then highly radioactive fission products could be dispersed along with the disintegrating machinery surrounding the steam break. This secondary effect, beyond the danger of flying metal and masonry, demands special attention to the possibility of a steam explosion in a nuclear power plant.

The first protection against the release of radioactive fission products is the ceramic fuel itself. The second barrier is the zircaloy fuel cladding, the third is the thick steel reactor vessel, and the final is the reactor containment building, as required by federal government regulation 10 CFR 50.55a. This building, which completely encloses the nuclear reactor and its associated steam equipment, is built of reinforced concrete, steel, or both concrete and steel. It is shaped as a square block, a vertically mounted cylinder, or a sphere. In any emergency, the containment building can withstand steam pressure in the range of 60 to 200 pounds per square inch (410 to 1,400 kp), and its intent is to prevent radioactive material from escaping into the surrounding area if the worst should happen.

A newer federal regulation now requires a commercial reactor containment structure to withstand the crash of an airplane directly into it. In 1988, the Sandia National Laboratories in New Mexico conducted tests to find if this requirement was possible to follow. A military jet fighter plane going 481 miles per hour (775 km/hr) was flown directly into a simulated concrete containment building wall. It left a gouge in the concrete only 2.5 inches (6.4 cm) deep.

As an unscheduled test of commercial reactor containment buildings, the Turkey Point Nuclear Generating Station in Homestead, Florida, sustained a direct hit from Category 5 Hurricane Andrew in 1992. Turkey Point has both nuclear and coal-fired generators. The coal-fired portion of the plant sustained $90 million in damages to a water tank and a smokestack. The nuclear portion of the plant reported no damage.

The intention of a containment building is to give a sudden, major steam leak or explosion a space to expand into, condense into water, and not escape into the outside air. The building, which is airtight, can be a

Three Mile Island-2 Containment Building

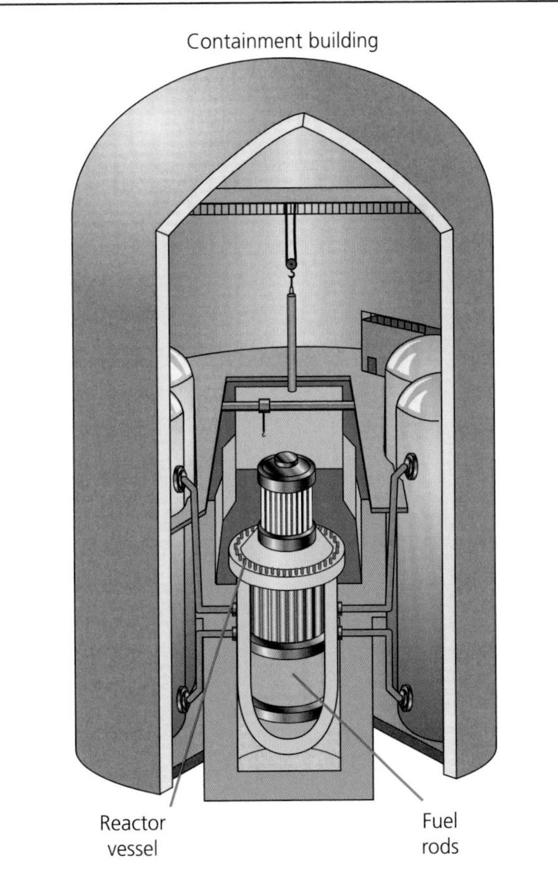

This cutaway drawing of the containment building of the Three Mile Island number 2 reactor in Harrisburg, Pennsylvania, shows the thick, concrete shell that protects the outside environment from radiation escape anywhere in the process of making nuclear power.

large, dry space, a vacuum, or an ice condenser. The ice condenser consists of refrigeration coils attached to interior walls of the building, designed to quickly cool the escaping steam without a need for a large expansion space, as is required for a dry containment. A vacuum building, with an added water-spray system, is used in many Canadian-built CANDU reactor systems.

A PWR containment building as just described is typically 10 times larger than a BWR containment. The BWR containment is usually a blockhouse, whereas a PWR containment is shaped like a soft drink can. The most recently designed containment structures are a combination of sphere and cylinder, being a can-shaped building with a half-spherical top. German reactor buildings are typically spherical. The only commercial nuclear power plants that have no airtight containment buildings are the now obsolete Russian RBMK reactors, of which the Chernobyl plant in the Ukraine was an example. The lack of containment was definitely a contributing factor in the disastrous leakage of fission products from the plant in 1986.

A cutaway painting of a European pressurized reactor containment building—this structure is double-walled and is meant to keep inside activity from reaching the outside and from outside threats from reaching the inside. *(© AREVA, Image & Process/NRC)*

A traditional BWR containment structure is a steel tank, or dry well, built around the steel reactor vessel. The dry well is housed in a cube-shaped building, constructed to withstand hurricane winds. If the worst should happen and a pipe suddenly breaks in the BWR cooling system, the steam expands into the dry well. Vent pipes at the bottom of the dry well direct the pressurized steam to a wet well below. The wet well is shaped as a torus or donut and half-filled with cooling water. The steam bubbles up through the water, losing its explosive energy. The secondary building around the containment structure is kept at subatmospheric pressure, so that no contaminated steam or air can escape to the outside.

All steam lines coming out of the reactor containment building are equipped with isolation valves that automatically slam shut in the event of an emergency. Every 15 years, the airtight nature of the containment must be confirmed with a leak test under the mandatory provisions of 10 CFR Appendix J to Part 50.

BACKUP GENERATORS

Under normal circumstances, a nuclear power plant provides its own operating electricity, which is a minimum of six megawatts of power. This power is used for everything from lighting the control room to running the coolant pumps. In the event of a reactor shutdown, either scheduled or emergency, these services must still be provided. It is essential that the coolant pumps remain in operation even after the fission process has been shut down, as delayed fission effects give a sustained power output of less than 1 percent of the reactor's rated production. One percent of a billion watts is 10 million watts, which causes a great deal of heat.

If a plant is shut down, it can still depend on the power network for electricity, consuming it instead of providing it, but there is a chance that the entire network has gone down, and the plant is on its own. All emergency situations, regardless of how improbable, must be accounted for, and the possibility of no electrical power in a power plant is managed using backup generators.

Each reactor in a nuclear power plant is provided with two diesel engines, connected directly to generators. In the event of power loss, the engines are automatically started, and the generators are switched into the plant power system. The typical nuclear plant backup engine is a Fairbanks Morse diesel, 12 to 18 cylinders in a V arrangement, originally designed for use in a tugboat or a train locomotive. The backup generators must be tested at least once per year.

Failure to start the backup generator during a test can shut down an entire power station until the problem is found and corrected. On December 14, 2009, a backup unit failed to start during a test at the San Onofre Nuclear Power Station, and the associated reactor had to be shut down for an entire weekend. At 13 cents per kilowatt hour, Southern California Edison lost about $130,000 per hour with this reactor out of service. For this reason and for the sake of ultimate safety, effort is expended to maintain the backup systems in good working order at nuclear power plants.

STEAM TURBINES

The steam turbine, as has been used for electrical power generation for more than 100 years, is an ideal machine with exactly one moving part and an efficiency of up to 90 percent. It sits horizontally on a floor, or turbine deck, with its massive driveshaft connected directly to an electrical generator. Mounted on the driveshaft are typically 10 wheels, each having on its periphery a number of vanes or buckets. The entire machine is enclosed in a close-fitting metal case, designed to provide bearings to support the shaft but not touching any wheel and sealing it against steam pressure.

Dry high-pressure steam is introduced through a pipe at the far end of the turbine case. It is directed at the vanes in the first wheel and causes it to turn much like a windmill. The steam, having expended some of its energy on turning the first wheel, then hits the second wheel, giving up more energy to the turbine shaft, and so on, down the line of wheels, until the steam is spent. This type of machine is used in large power plants, coal and nuclear, and called a condensing turbine. By the time it reaches the end of the wheel line, the steam pressure is reduced to below atmospheric pressure.

The steam is directed out the end of the turbine to the condenser, below the deck and under the turbine, where it is condensed back into liquid water by an exchange of heat with the externally connected cooling tower.

Plants of modest power, below 100 megawatts, run the generator at a rotational speed of 3,600 RPM (revolutions per minute). Large nuclear power plants use a four-pole generator and run the turbine at half that speed, or 1,800 RPM. The shaft speed is maintained precisely, to ensure connection with the larger power network, on which electricity alternates between positive and negative exactly 60 times per second.

At the end of the driveshaft, opposite that of the generator, is one small extra turbine wheel. It is a driven wheel and does not contribute to the

Pipes deliver steam from the reactor to four turbines at the Bruce B Nuclear Power Plant in Tiverton, Ontario, Canada. *(Norm Betts/Bloomberg via Getty Images)*

power on the shaft. It is the oil pump, providing pressurized oil to the bearings in the turbine in a recirculating system. Oil is constantly fed to the bearings, where it eventually falls though holes in the bottom of the bearing surfaces, is collected in a sump, and is forced back into the bearings by the turbopump.

At the very end of the turbine shaft, outside the turbine case, is the barring engine. Its purpose is to keep the turbine turning at a low speed, even if the reactor is completely shut down. If the plant is down for upgrades, refueling, and maintenance, the turbine can be idle for months, but still it must turn. If allowed to sit in one position, the heavy wheels and shaft would begin to sag. Or, if the wheels were idle just as steam was being introduced slowly in a plant start-up, one spot of the wheels would be heated while the rest remains cool, and this would cause warping. For these reasons the barring engine keeps the turbine rolling under all conditions.

GENERATORS

The generator in a nuclear power plant is a rotating machine producing alternating electrical current, the frequency of which is completely dependent on the speed of the turning shaft. The concept is similar to the 12-volt alternator in an automobile. A major difference is that an alternator produces about 700 watts, and a nuclear power plant generator produces 1 billion watts.

Both the alternator and the power plant generator use a rotor, turning on the shaft, and a stator, standing still and surrounding the rotor. Both the rotor and the stator are wound with copper wire. The windings in the stator are used to produce a magnetic field, which induces electrical power in the rotor when it turns. The power is picked off the rotor using slip rings and graphite brushes. The generator produces electricity alternating as a perfect sine wave, at a typical maximum of 24,000 volts. The electricity is produced in three sine waves on three wires, all alternating at the same frequency but in three phases, A, B, and C. Phase B lags phase A by 120 degrees, or one-third of a complete cycle. Phase C lags phase B by 120 degrees.

A nuclear plant generator is as big as a mobile home, and its energy efficiency is an astonishing 99 percent. All but 1 percent of the energy on the driveshaft is converted directly into electricity. That remaining 1 percent, or 10 million watts, is wasted as heat in the copper windings and is enough to melt the entire machine in short order. This overheating is prevented, and the machine is brought to full efficiency by two cooling schemes. The stator is cooled with water, and the rotor is cooled with hydrogen gas, both operating in separate closed loop systems.

To cool the copper windings, the wires are made hollow, and coolant is pumped through them continuously. For maximum cooling efficiency, hydrogen gas with its excellent thermal conductivity is used in the confined space of the rotor, and the entire generator is internally pressurized to 75 pounds per square inch (517 kilopascals) with hydrogen. It is very efficient for the rotor to turn in an atmosphere of hydrogen rather than air, because hydrogen imposes a lower drag than air. The hydrogen and the water coolant temperatures, flow rates, pressures, and

This is the generator end of the turbogenerator on the turbine deck at a nuclear power plant. *(NRC File Photo)*

hydrogen purity are all monitored continuously in the control room. The two coolants are cycled through heat exchangers to maintain low temperatures.

THE SWITCH AND TRANSFORMER YARD

The electricity from the generator leaves the turbine building through three copper conductors. A nuclear power station is usually hundreds of miles from the ultimate electricity consumers, and the product must be transported to them with a minimum of loss. This is accomplished by increasing the voltage of the electricity for long-distance transmission. Very high-voltage electricity, carried on wires held high in the air by steel towers, can travel for far distances and then be down-converted to lower voltage for use in factories and homes. Depending on the circumstances, the voltage can be increased to as high as 500,000 volts.

Conditioning of the electricity for transmission is accomplished in the switch and transformer yard, located on the power plant property close to the turbine building. Transformers are used to increase the voltage of the electricity, using a magnetic induction principle discovered by the English physicist Michael Faraday (1791–1867) in 1831. Faraday found that a magnetic field produced by electrical current in a coil of wire induced a current in another coil of wire if the two were magnetically coupled together by a common iron core. The voltage of the induced current could be changed at will, by varying the number of windings in the induced, or secondary, coil. If, for example, there were twice as many coil windings in the secondary coil as in the first, or primary, coil, then the voltage was doubled. This principle is used extensively in electrical power transmission, stepping the voltage up and down as the electricity generated in a power plant is transmitted, spread out, and distributed to consumers.

The power plant transformers, taking electricity directly from the generator, must be oil-cooled to prevent overheating, using the same closed-circuit philosophy used in several places in the design of the power plant. Cool oil is pumped into a sealed case surrounding the transformer, and the heated oil is then piped out the top of the transformer and to a heat exchanger, where it is recooled and reused. Separate transformers are used for each of the three phases of alternating current from the generator.

A second function of the high-voltage equipment outside the turbine building is switching and circuit protection. The circuitry and the generator are protected from a massive short in the high-voltage transmission lines by circuit breakers, set to interrupt the circuit automatically on a preset detection of an unusual load. A sudden short circuit, probably caused by a transmission tower being blown down, is detected as an inappropriately high current. In a split second, a relay trips, a solenoid is actuated, and a set of electrical contacts fly apart. In the power plant yard, the voltage is exceptionally high, so the contacts must be separated widely to prevent the electricity from jumping the distance as an arc.

All that is necessary to fully disconnect the circuit is distance between the contacts, but the contacts can only move apart at a finite speed, and, as they are being flung open, an arc always develops between them. Once the contacts are wide apart, a fat, blue, continuous spark

A portion of the switching yard at the Bradwell Nuclear Power Station in Essex, England *(Justin Kase z07z/Alamy)*

through the air maintains an electrical connection. To correct this, "air-blast" circuit breakers are used in power plant yards. As soon as the contacts are fully opened, a sudden blast of compressed air across the arc literally blows it out like a birthday candle, and the circuit is finally open. This same strategy is used for automatic circuit-protection breakers and for the switches used to turn on the power and send it out into the consuming world.

These are the basic parts of a nuclear power generating station. The next chapter focuses on a tour through an operating power plant, making electricity from nothing more than the splitting apart of unusually heavy atoms.

7 A Walk through a Nuclear Power Plant

All the components, subsystems, and assemblies explained in the previous chapters add up to a nuclear power station, connected to the greater power grid and contributing to the world's sum of consumed electrical power. The plant is complicated, and as a whole it is somewhat overwhelming. In this chapter, to bring this large grouping of machinery into scale, the reader will experience a walk-through of a nuclear power plant. Most plants consist of two nuclear reactors and the associated turbogenerator equipment, with each unit acting independently to produce power. In the United States, some plants have one reactor and a few have three. In Europe, there are plants with four reactors on the same site.

There can be hundreds of people working at a power plant, doing everything from polishing floors in the hallways to directing the operation of the plant. In a maintenance outage, in which one or both reactors are shut down for refueling, the load of personnel is greater, as specialists swarm over the equipment to make small repairs or replace entire systems with updated machinery in the days of stillness as the fuel is carefully, slowly shuffled in and out of the reactor cores. Pipe fitters, computer programmers, mission planners, radiation workers, engineers, and electricians take up space in the parking lot and cause walking traffic jams in doorways.

All but one of more than 400 nuclear power plants in the world are built near flowing water, for use in the cooling towers. The Palo Verde

Nuclear Generating Station near Wintersburg, Arizona, is unique in that it uses treated sewage from several nearby cities as a water source for its six cooling towers. Generating 3.2 billion watts of electrical power, it is the largest nuclear power plant in the United States. In general, a nuclear power station is located on or near the banks of a river.

Nuclear plants are also usually located on flat land. This cuts down on the amount of earthwork that must be performed. A nuclear plant must be located near major transportation routes, so that heavy equipment and lead-shielded fuel can be moved in and out of the site. The reactor vessel is probably the heaviest, most ungainly component, and it must be shipped to the site on a river barge or on a special wide-load tractor trailer. Many tons of steel must be moved to the construction site, as well as thousands of truckloads of mixed concrete. There must be a concrete plant within economical distance of the building site, and for this reason building contractors have been known to first build a new concrete plant before beginning construction of the power plant. In an arid location such as Arizona, a long string of water trucks must be employed, to have something to mix with the concrete dust.

A stringent and difficult requirement for approval of a nuclear plant building site is that it must have minimal seismic history or a low predictable chance of a seismic event. If a plant is built in a region with seismic history, then special precautions must be employed in the design to prevent damage from ground movement, and this adds significantly to the building expense. The fear is that an earthquake could break open the primary cooling loop and spread fission products from damaged fuel. A thorough geological examination of the site and its surroundings is necessary. Evaluation of seismic probabilities is difficult, and there is hardly a plot of ground on Earth that has no geological faults underneath it, as the tectonic plates of the continents have been wandering for billions of years. As best as can be determined, nuclear power plants are usually located on ground with no known seismic activity.

It is a costly enterprise to build a nuclear plant. At current prices, the cost is about $4 billion per reactor. With a private investment of this breathtaking size, it is best to build the plant to last as long as possible. Nothing flimsy is used in nuclear construction. Everything, from hinges on doors to cables running signals to the control room, is of heavy-duty specification.

The grounds of a plant are uniformly covered with a layer of gravel, meant to encourage water drainage during rainstorms while preventing vegetation from growing. It is not a garden, an animal sanctuary, nor a lovely spot to attract tourists. Stand still and listen. The entire place seems to hum at 60 hertz, and sometimes there is a crackling, almost popping sound, as high-voltage electricity leaks from bare metal cables into the air. It is the heaviest of heavy industries, and it exists for only one purpose, to push electrons through three copper wires.

THREADING THROUGH THE SECURITY MAZE

The first thing to notice about a nuclear power plant upon approach down the access highway at night is the lighting. Every corner of the grounds, the buildings, the roads, and the massive switch and transformer yard are lighted from every angle, leaving nothing in the dark. There is no place to hide outside a nuclear plant. The light is strong enough to wash out the blue glow of ionizing air nitrogen on the high-voltage equipment, but

The security gate at Three Mile Island Nuclear Power Plant in Harrisburg, Pennsylvania—the guards are armed. *(AP Images)*

under the right conditions of humidity it is visible on the triad of high wires leaving the plant and spanning across the highway. It resembles St. Elmo's fire, the eerie illumination of the rigging on a sailing ship due to an atmospheric electrical charge.

The grounds are surrounded by a high chain-link fence with barbed or razor wire on top, laced with intrusion-detection equipment. Armed patrols, often accompanied by trained dogs, walk up and down the lines, all day and all night. There has always been a fear, even before there were organized terrorist attacks on the United States, of sabotage or mischief to the very large fixed investment of a plant. Without intensive 24-hour security, no insurance company would underwrite a power plant.

There may be a visitors' center outside the gate. Inside is a museum-like setup, showing photos of the plant being built, the head of the power company at the ground-breaking ceremony, wearing a hard hat and smiling as he throws a shovel of dirt at the camera. There are brightly colored engineering models of the reactor, the steam equipment, and the turbine, left over from the sales effort by the manufacturer. There are brochures containing specifications and statistics of the plant and its power source, the nuclear reactor.

There is a guardhouse at the only gate into the fenced area. Other areas of the fence can be opened for special needs, but for all day-to-day operations including fuel transport everything has to go through this one bottleneck. To gain entrance past the guard, one must have an identification badge or some sort of authorization from higher powers in the power company. The car or truck is waved through, and it may proceed at a very slow, prominently posted speed limit. Exceeding this speed will attract attention to the vehicle and could result in gunfire if the guards are sufficiently alarmed. The guards can become bored with the same lack of action day after day, and care must be taken not to do anything that they may find exciting or monotony-breaking. As a rule, they take their jobs, to prevent unwarranted access to the grounds, very seriously.

There is plenty of parking available at a nuclear plant. Normally, the lot is about 1 percent full. During outages, when the maintenance is performed and the upgrades are installed, the lots can overflow, as outside workers flood in. It seems a long walk to the entrance portal, but all distances at the plant are large. The plant is spread out over many acres. Nothing is crowded. There is plenty of room for everything. Most plants, in fact, are built with enough extra room to install another full-sized generating station on the site. Finding, licensing, and preparing a building site are time consuming and costly. It can take 10 years to obtain a building license, so a site is precious, and buying the land is the least expense. A utility makes the plant site as big as possible. So, it is a long walk from the parking lot.

There is only one entrance portal for personnel. There can be other breaks in the buildings, but they are opened and used only for the installation or removal of large components, and such operations are served by special security detachments. The nuclear power plant is not just one building, but is a virtual town of buildings, connected by hallways, underground tunnels, or covered walkways. Personnel movement is well controlled, making it difficult to wander into an unsafe area near high voltage, high radiation, or other risk.

Entrance security has been changed and improved at industrial plants over the past few decades, as threats have increased in variety and severity. Now, entering a nuclear plant is very similar to the ordeal of going through airport security. Metal detectors are used to make certain that no weapons are being brought in, and pockets must be emptied into trays. Some plants may even require the removal of shoes. Each person entering is examined, right down to the belt buckle.

This security pat-down wins entrance to the noncritical or safety-related rooms and buildings on the site. This access includes offices, storerooms, restrooms, classrooms, briefing or conference rooms, the cafeteria, the library, the infirmary, and the remote emergency center. This last room is reserved as a redoubt, where operators and engineers can gather if there is an extreme emergency in which access to the power complex, including the control room, is blocked because of radiation spills or severe damage to the reactor structure and the turbine building. It is equipped as a subset of the control room, with all critical indications and readouts from the control room instruments wired into the room. It is not possible to control the plant from this room or from any other room on or off site, but the moment-to-moment status of the plant and its systems can be evaluated from this room.

Of even greater importance and constant use in this building complex is the control room simulator. This is a room that is a direct, perfect copy of the control room, having all the hundreds of switches, indicator lights, annunciators, and meters on the vertical panels and horizontal benches that are in the real control room. This one, however, is not connected to a power production system. It is connected to a computer cluster in an adjacent room, and this arrangement is used to digitally simulate everything about the actions of a nuclear reactor operation. It is used for operator training, retraining, testing, evaluation, and updating to the latest added or subtracted equipment from the plant.

The simulator is particularly adept at challenging operators with faked equipment malfunctions, accidents, or complete disasters. During a training session, it can throw in an equipment failure as small as a lightbulb burnout in a panel indicator or as grave as a complete loss of coolant in the reactor vessel. Using this facility as a training tool, the operators can experience and learn to deal with problems that are so rare they are known only in theory. An operator must think fast and think clearly in an emergency and sometimes the best action is to do nothing at all. The operators are conditioned to know how to act in a vast range of emergency conditions from many hours in this room. The experience is so real, it can cause a seasoned senior reactor operator to break into a sweat and experience heart palpitations under simulated conditions.

Everyone, not just the reactor operators, requires a semiconstant training schedule in the nuclear plant, and for this action classrooms are provided. New employees, old employees, and contract or temporary employees are first taught how to avoid radiation and why it must be avoided. Safety is the first concern of the power company, and the safe uses of the facilities and the equipment at the nuclear power station are drilled into everyone.

WHOLE-BODY RADIATION MEASUREMENT

If a person is qualified and requires access to the working part of the power plant, then he or she must pass through another choke-point portal. Beyond the office complex of the plant, a hardhat must be worn as protection against hard objects dropped from above or a slip and fall against the concrete. There is no eating or drinking beyond this portal, and although the plant has been designed to be as safe as possible, it is not lined with pillows. The power plant is made to generate a great deal of power and take a lot of stress. The machinery to do this is heavy and hard with moving parts, high voltage, and intense, invisible radiation. There are a thousand ways to get killed being near it. One must be careful and constantly mindful of what is happening.

The purpose of this portal is to count radiation on or in the human body. Personnel coming into the working end of the plant are radiation counted, and people going out of the plant are counted. The goal is to have no radioactive material carried into or out of the plant. This prevents the spread of radioactive contamination from a worker's shoes, pants legs, or hands

A radiation counter checks a worker for excess radiation before he enters the control area at a nuclear power plant. *(NRC File Photo)*

outside the inner area. It also prevents any externally picked-up radioactive material from contaminating inner walkways, walls, or doorframes. Any radiation from the outside would confuse the monitoring of radiation produced in the plant.

The radiation is literally counted, usually against a set period of time on a clock. Individual radiation particles are detected and summed in a register, and at the end of the counting period the resulting number is evaluated. Everyone is slightly radioactive, but if a person is contaminated with radiation causing an unusual count, then further evaluation is necessary.

There are many hand and foot monitors stationed around the inner plant, in all places where touching a surface could possibly leave a trace of contaminant. These are radiation rate meters, equipped with an alarm setting for high values. They are usually set to measure gamma and *beta rays. Alpha particles* are an important radiation mode, but they are difficult to measure, as they have very little penetration power. Alpha contamination is often masked by overlying material acting as a shield, and detection of an ingested alpha source is particularly challenging. Handheld alpha monitors employ a special Geiger counter probe having an extremely thin detection window made of Mylar film. It can detect a naked alpha particle source, but it must be held very closely to the radiating material. An alpha-detector probe held in the palm of the hand can detect a contaminating alpha source.

Constant monitoring and checking of general radiation levels and spots of radiation in the plant, as well as maintenance of the radiation detection equipment, are handled by the health physicists (HPs). Their jobs never end, continuing in importance when the plant is shut down for refueling and even into the future when the plant is permanently shut down and demolished. With handheld radiation detection equipment and

A whole-body counter doing a complete survey of the radiation emitted by a person's body in Zagreb, Croatia *(AP Images)*

built-in monitor stations throughout the plant, they monitor every action by personnel involving radiation, including the complicated refueling operations and steam pipe repairs.

The large counting machine in the entrance portal is a whole-body counter, designed to integrate the radiation dosage measured over the entire person, from the top of the hardhat to the soles of the steel-toed boots. Each person entering the controlled area of the plant must stand in the doorway-shaped device and be counted on top, bottom, and both sides. It is not the ultimate whole-body counter, in that it makes a cursory evaluation by only counting for about 30 seconds. This measurement, however, is sufficient to detect surface contamination being brought into or out of the working section of the plant. The true whole-body counter is over in the infirmary. There a person lies horizontally in half of a body-sized detector, with a second half lowered from overhead. It can take an hour to be counted on this machine, but it will find anything with beta or gamma radiation that the subject has eaten, breathed, or rubbed against.

Everyone in the plant must have a whole-body count upon starting work in the nuclear industry. This gives a beginning or starting point to a monitored history of radiation exposure for each employee. Every medical and dental X-ray, CAT scan, or other medical diagnostic procedure involving radiation is noted, and the daily radiation exposure of the person is recorded. The total yearly radiation dose per person is recorded in the health physics office, to make certain that a worker does not receive exposure that would affect the body. All personnel's radiation exposure must stay within the *Nuclear Regulatory Commission*'s published limits. This policy applies not only to maintenance workers laboring next to steam pipes, but to everyone on the plant property, including the security guards at the fence.

THE CONTROL ROOM

The control room, where all systems in the power plant are monitored and manipulated, is usually located on the uppermost floor in the building complex. Although its position would offer a splendid view of the turbine deck, the cooling towers, and the surrounding countryside, the control room has no windows. The recessed overhead lighting gives no clue as to the time of day or the season of the year. In the control room, the lights are always on, and there is a constant stillness. The only evidence that power is being generated at this location are the positions of indicating needles on gauges.

The control room is the nerve center of the power plant, and it is a nexus of wires and cables from the far corners of the plant, all bringing information into the room or transmitting electronic commands. The room directly below the control room is the spreading room, which is a space as large as the control room that is used as a meeting point for hundreds of cables. They are arranged neatly, arriving horizontally through the walls of the room in trays or conduits, with each cable labeled and routed to its appropriate station under one of the control panels. A cable is addressed to its location at a specific point at a control panel overhead, on the control room floor, where it makes a 90-degree turn straight up and disappears through a hole in the ceiling. In the control room, each cable comes up through a hole in the floor under a control panel and is routed to its switch, meter, or lighted indicator.

The control panel sections, with vertical displays and horizontal benches, are often arrayed as octagonal walls of a room, with desks for the senior reactor operator and the control room supervisor at the center of the space. These control panels are false walls, and behind them are more vertical panels, containing more indicators and gauges, monitoring less critical functions, such as fixed instruments looking for radiation leaks in cooling systems.

A nuclear power plant control room is probably safer than a doctor's waiting room, as no one shows up for work sick in this occupation. However, just to be certain of personnel safety under the worst conceivable conditions, the control room is surrounded by blast and radiation shielding. In the very unlikely event of a steam explosion on the turbine deck, there is sufficient armor in the walls, floor, and ceiling to prevent shrapnel from a disintegrating turbine or radiation from a reactor vessel rupture to affect the calm atmosphere of the control room. There is a chance of a fire whenever this much electrical equipment is concentrated in one spot, but the probability is minimized by the lack of burnable material in the room, the use of nonflammable insulation in the wires, and the presence of automatic and manual carbon-dioxide fire extinguishers.

As a work environment, the nuclear power plant control room is near ideal, with its neat, logical layout and its businesslike atmosphere. There is no eating or drinking in this room, access is tightly controlled, and no visitor is allowed to lean over the controls. Where a nonoperator can stand is marked off on the floor. Just stand there, quietly, and watch the lights.

STANDING ON THE CORE PLATE IN A BOILING WATER REACTOR

The largest, heaviest, and possibly costliest single component in a nuclear power plant is the reactor vessel. It is a large, cylindrical steel tank, at least eight inches (20 cm) thick, with rounded, half-spherical ends. Except for the removable top, it is one continuous piece of steel, forged and then machined with great precision. It is static-tested for any tendency to leak under high internal pressure. After successful testing, it is transported to the building site on a special low wide-load tractor trailer or a river barge.

The skill and the facilities necessary to build a reactor vessel are becoming scarce. There is no longer a steel plant in the United States that is capable of making one of these vessels, as there has not been an order for one in the past 30 years. There may be one company that can still fabricate these units, Toshiba Power Systems of Tokyo, Japan. If pushed, Toshiba can probably turn out four reactor vessels per year. This will not meet demand if worldwide orders for nuclear reactors increase in the coming decades, as is expected, and the steel industry will have to retool to supply builders with this essential component.

The size of the reactor vessel depends on the type of reactor. A typical Westinghouse PWR vessel has an internal diameter of at least 132 inches (335 cm). A large example can be 176 inches (447 cm) wide, not including the thickness of the walls. A BWR vessel is at least twice this size.

The concrete reactor containment structure is built around the reactor vessel, and the complex tangle of pipes that conduct cooling water into and out of the vessel bind it into the structure of the building. Most components in the power plant can be replaced if necessary, but not the reactor vessel. It is too deeply embedded in the plant. The working lifetime of a nuclear power plant thus depends on the life of this one piece. When the vessel no longer passes safety tests, the plant is permanently shut down.

These vessels are well built of stout material, but there is a failure mode that bears watching. Steel can become brittle and prone to breakage if hydrogen atoms become lodged in the crystal structure of the material. In a water-cooled nuclear plant, there are two ways that the steel vessel can become contaminated with hydrogen. It can be derived from the water in the vessel by radiolysis, which is the breakdown of water under intense radiation. Individual hydrogen atoms also materialize within the steel structure, as neutrons leaking out of the reactor core decay into protons, or hydrogen nuclei. A neutron set free of a nucleus has a half-life of only 10.6 minutes, and it tends to move slowly through the steel.

REACTOR START-UP

A nuclear plant has five or six teams of reactor operators, with each team typically consisting of three reactor operators plus a senior reactor operator. They work in eight-hour shifts, with the control room fully manned 24 hours a day. After a refueling outage, an extended maintenance outage, or when the first fuel has been installed in a new nuclear power plant, after every system has been tested and checked, it is time to start the reactor. This is a complicated procedure, and a full description would require an entire book, but for a typical nuclear power generating station the start-up can be described in the following five parts:

- Bringing the reactor from cold shutdown to criticality. In a state of shutdown, fissions in the reactor occur randomly, at a vanishingly low rate, and no power is being produced. When the reactor is brought to critical condition, fission begets fission to the point where the reactions are self-sustaining. At criticality, exactly the same number of neutrons are lost as the number that are produced by nuclear fission in the uranium-235 component of the reactor fuel. A critical condition, however, does not imply the production of power. A reactor can be critical at essentially zero power.
- Reactor heat-up. The nuclear reactor, and all of its attending coolant services, must be brought up to operating temperature slowly and carefully. The start-up crew must be careful not to stress anything by a sudden temperature increase, and they need time to check and recheck everything to make certain that all systems are capable of operating at the extreme end of the power range. At full power, temperatures, pressures, and flow-rates subject the power system to a great deal of stress, and it is necessary at this point to verify that all pumps, valves, and pipes are in good condition.
- Turbine start-up. The reactor is now operating at low power, and the steam line to the turbine opens. The turbine starts rolling. The generator is synchronized into the electrical network. It is not contributing much to the network, but it is connected.
- Raising the coolant flow through the core. Everything seems to be operating smoothly, so the valves are opened, and power increases as the coolant flow-rate increases. The cooling towers start to blow steam out the tops as river water evaporates in the sprinkler system.

- Full power is attained. The power plant is now fully operational and is on automatic control. The operators continue to observe and make control corrections when necessary.

These steps are performed literally by the book, or the books. A set of reactor operating manuals is always in the control room, neatly arranged on shelves near the center of the room.

At the center of the array of control benches is the reactor mode switch. It is a handle-shaped switch, with three positions: START-UP, RUN, and SHUT DOWN. This switch is moved to the START-UP position. For start-up, a reactor must have an installed artificial neutron source. It is usually a small bit of a radioactive alpha ray emitter mixed with beryllium mounted at the end of a motorized metal pole, and its purpose is to guarantee a source of neutrons even if there is no fission in the reactor. This puts a known signal level into the neutron detection equipment as a check before start-up. All chain reactions must start with a single neutron, and this source ensures that the initiating neutron exists. The start-up begins with attention to the source range monitor, which is a neutron detector attuned to the low levels of a zero-power criticality situation.

The reactor is taken from cold shutdown to criticality by gradually pulling out the neutron-poison control rods. Only one rod can be moved at a time, and there are hundreds of them, arranged in a matrix. Each rod is given a number consisting of its X-Y position in the matrix, and a map of the rods showing individual positions is provided on one of the vertical control panels. A rod can be moved a small increment by pushing a button. It is possible to move a rod continuously by holding down the button. Control rod withdrawal is accomplished in accordance with the rod sequence check-off sheet. Even in a world of computers, the operators must check off individual control rods on sheets of prepared paper according to a prearranged schedule.

The rate of change of neutron activity in a neutron chain reaction is the reactor period. This is the time, in seconds, required for the neutron level to change by a factor of approximately 2.72. This number is the root of the natural logarithm, or "e." A negative period means that the neutron rate is going down, and a positive rate means that it is going up. When approaching criticality in a power reactor, the period should be kept between 30 and 100 seconds. At full shutdown, at criticality, and at full power, the rate is exactly zero.

A wide view from inside the control room at a nuclear power station: A cold start-up is not taking place here. *(NRC File Photo)*

When the source range monitor approaches full scale, the intermediate range monitor takes over the task of indicating the neutron level in the reactor, and the source and the source range monitor detector are withdrawn from the core. Watching the period instrument and the neutron detector, an experienced operator can see that the reactor has attained criticality, and the control rods are pushed far enough back into the core to maintain a steady neutron level and a zero period. The reactor is now critical at a negligible power level.

The temperature of the reactor is now raised carefully, at a rate of less than 500 degrees per hour. To increase the power level, and therefore the temperature, the reactor must be temporarily taken out of critical condition and made mildly supercritical. Controls are pulled out of the core slightly. The power level will rise slowly to about 10 percent of the full power rating, at which time the reactor is pulled back down to exact criticality by inward movement of the control rods. The water temperature in the reactor vessel increases to above the boiling point.

The plant is now ready to roll the turbine. The intermediate range monitor is withdrawn from the core and the average power range monitor takes over, indicating the increased neutron level in the reactor. Controls are again withdrawn slightly to bring the reactor power slowly and smoothly up to 40 percent. When the 40 percent power level has been reached, the controls are returned to the position of exact criticality, and the period meter reads infinity. The reactor mode switch is turned to the RUN position.

There is now enough steam being generated to automatically open the turbine bypass valve, and the turbine begins spinning at 900 revolutions per minute (RPM). It must be allowed to roll at this slow speed for two minutes, to allow the temperature to equalize in the periphery of the turbine wheels. It would be disastrous to have one side of the turbine hotter than the other side, as the metal would expand asymmetrically.

After the warm-up, the turbine is brought up to the full speed of 1,800 RPM. At this point, attention turns to the turbine control panel. There are two buttons, FAST and SLOW. Pushing the FAST button will slightly increase the turbine speed, and the SLOW button will slightly decrease it, opening and closing the steam valve by small increments. At the center of the turbine control panel is a most interesting instrument, named the synchroscope.

The synchroscope is a round dial, marked off in degrees of a circle. At the top is 360 or zero degrees, and at

Synchroscope

The synchroscope. When the needle is still and at the top, the power plant generator is running with its alternating current in sync with the rest of the power grid.

the bottom is 180 degrees. An indicating needle is pinned at the center of the dial, and it points to the difference between the phase of the sine wave, alternating current being produced by the generator and the sine wave existing on the outside power network. If the needle is straight up, it means that the two power sources, the generator and the rest of the world, are in perfect sync. If it is at the bottom of the dial, it means that the two are exactly out of sync. To correctly add power into the network, the generator must be perfectly synchronized with the existing alternating current of the network. At a turning rate of exactly 1,800 RPM, the generator is producing 60 hertz electricity, but it is not necessarily in sync.

Putting the generator in sync with the power network is an entirely manual operation and requires concentration. To come to synchronization, the synchroscope needle must be moved. To accomplish this, the turbine speed is adjusted slightly up by punching the FAST button. Now, not only are the generator and the network out of sync, they are not even oscillating at the same rate. The synchroscope needle begins to spin clockwise, slowly. Wait until it gets exactly to the top of the dial and then punch the generator output circuit breaker. The synchroscope freezes at 360-0 degrees. The generator is now connected to the power network, and the plant is producing electricity. Depending on the needs of the network, the generator load set-point control is then turned to a percentage of the plant's total power capability. The plant will now maintain this power production as if it were on cruise control.

The nuclear power plant is now operational. It is capable, on demand, of producing its maximum rated power.

FUEL MANAGEMENT

Fuel can be safely loaded, unloaded, and stored in the reactor building using a carefully designed system. New enriched uranium fuel delivered to the plant is radioactive, but not dangerously so. People can actually pick up an handful of uranium fuel pellets and examine them with no fear of harm. The fuel, which is mostly uranium-238 oxide, decays slowly, with a 4.5 billion year half-life, resulting in alpha and gamma radiation at a very low level. After 50 billion years, a fuel pellet will have become primarily lead-208 oxide.

Spent fuel, or fuel that has been used in the fission process, is another matter. About 5 percent of the uranium-235 will have been fissioned, and the fission product atoms will be roughly half the weight of the

The spent fuel pool at a nuclear power plant. It is filled with water, and this provides excellent radiation shielding and cooling as the fission products in the fuel decay away. *(NRC File Photo)*

original uranium atoms. The nuclei of these newly created atoms will be unnaturally heavy with neutrons and will seek to shed or convert the excess into protons. The result is intense and possibly dangerous radioactivity. Spent fuel must be treated carefully and with planning, as it is a deadly radiation source. Spent fuel is also hot, as delayed fissions are still taking place and the fuel is still generating power after the reactor has been completely shut down. The power level is very low and the delayed fissions decrease quickly after the shutdown, but they must still be taken into account. Measures must be taken to ensure that the spent fuel does not overheat.

About every two years, a nuclear power reactor must be refueled. To keep the workers safe from radiation as the core vessel is opened, the spent fuel is removed by remote control. Furthermore, the top of the reactor building is flooded with water. Moving spent fuel through deep water meets two requirements. Water is an excellent radiation shield, covering an immersed radiation source and stopping even the most energetic gamma rays. Water is also an ideal coolant, in constant motion around a hot fuel assembly by thermal convection. The top floor of a reactor building is made watertight, with a remote-controlled crane in

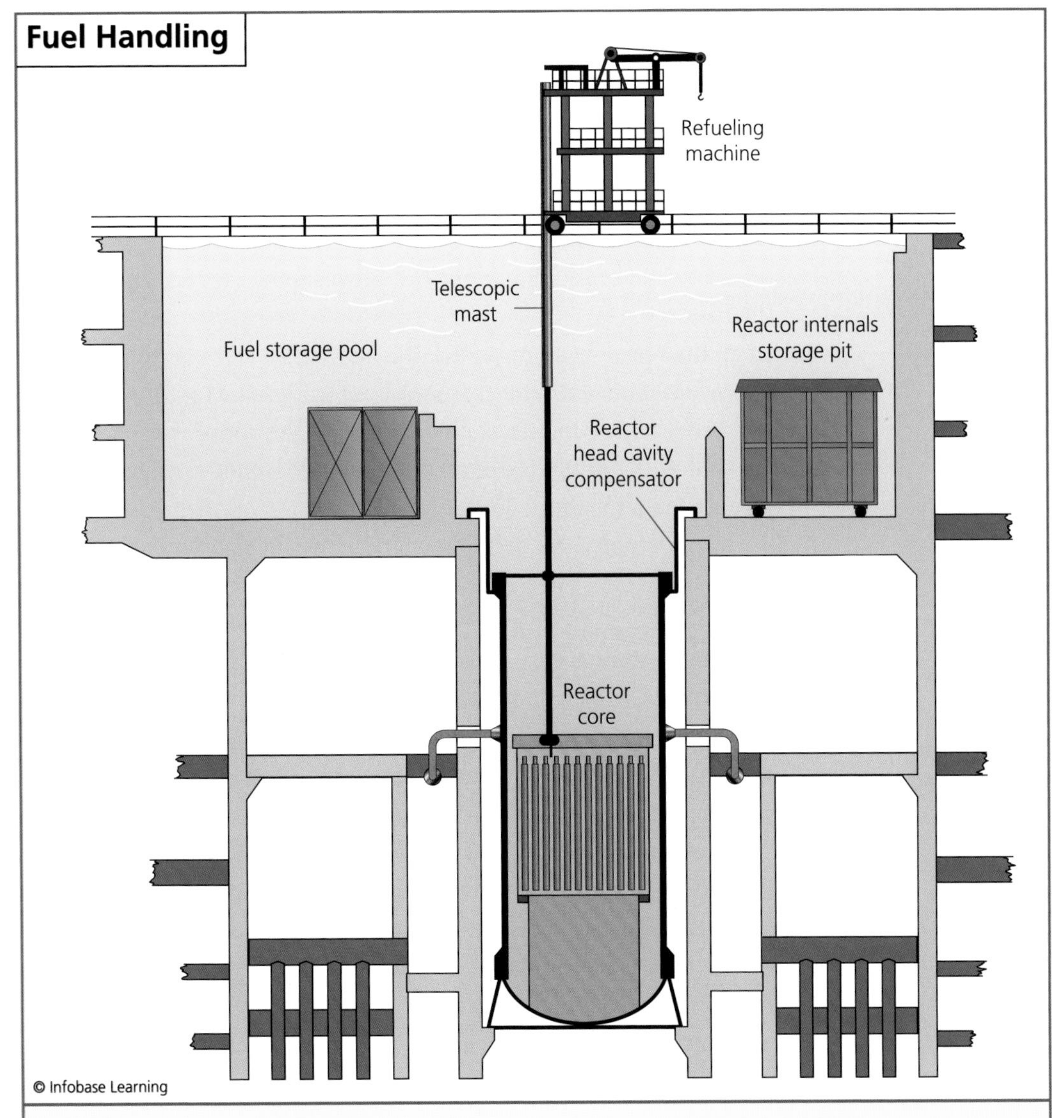

A schematic diagram showing how a nuclear power reactor is refueled. The fuel-handling machine runs back and forth on rails above the fuel pool, with a telescoping rod that reaches down into the core. The entire area is flooded with water, which makes a transparent radiation shield over the open reactor vessel.

the ceiling. PWR plants generally have a circular reactor building and are equipped with a polar crane. A polar crane is able to lift an object at any point on the top floor by moving as a radius and an angle on the ceiling. A BWR usually has a square reactor building, and a traveling crane is in the ceiling. A traveling crane addresses a point on the top floor by moving in two directions, in and out and side to side. These electrical cranes are manipulated from a glassed-off control room at one end of the building.

To refuel, the cap on the top of the containment structure is first lifted off by the crane and set aside in an assigned place. The hexagonal nuts holding on the reactor vessel tops are then spun off using a large power socket wrench and carried carefully to the sites of the holding studs atop the vessel using the crane. Each nut has an assigned storage space at the side of the room.

When all nuts have been removed, the hemispherical top of the reactor vessel is then lifted off with the crane and put aside. The top of the reactor vessel, still filled with coolant, is now exposed. The room is then flooded to a depth that will completely cover the fuel assemblies as they are lifted out of the core, usually about 30 feet (10 m).

A reactor is refueled by removing one-third of the fuel taken from the center of the core. Fuel burns out faster at the center, as the density of fission-inducing neutrons peaks at this point in the core. Fuel assemblies, looking like very tall boxes made of thin sheet metal, hold the long, thin fuel rods in a square, vertical matrix, not touching each other. A special tool on the general purpose crane grabs a handle on the top of each assembly, pulls a trigger, and releases the fuel assembly from its locked position in the core plate. Carefully, the assembly is lifted up clear of the top core-plate. Workers are careful not to hit a fuel assembly against anything solid, as it would come apart, spread radioactivity all over the top floor, and necessitate an expensive cleanup. The entire length of the fuel assembly is still under water in the flooded room. It is easy for the refueling technicians to see what they are doing through the clear water. The crane pulls the assembly to the side of the room, through water, to a rack in the storage pool, where it is lowered into its assigned position.

Once a third of the fuel has been removed from the core center, all the remaining fuel is moved, bringing it in to fill the empty locations. As a last procedure, the new fuel is removed from its shipping container and placed in the now empty locations around the periphery of the core. When the last new fuel assembly is locked in place in the reactor core, the floor is drained. The fuel pool remains full of water. The vessel top is picked up by

Technicians on the refueling crane above a BWR in California *(Tom Tracy Photography/Alamy)*

the crane and lowered gently onto the reactor vessel. The nuts are picked up one at a time and are screwed down on the top, and, as a last measure, they are tightened individually to a specified torque. The concrete cap is then placed back on the containment vessel, and the crane is parked at one end of the building. Refueling is complete.

In theory, the spent fuel should be allowed to cool for a specified time and then loaded into Holtec HI-STAR 100 metal transport casks, loaded onto trucks, and sent to the National Spent Fuel Repository. The HI-STAR is a large metal cylinder, painted white, with an impact limiter attached to both ends, making it resemble a giant dumbbell. Only one HI-STAR can fit on a flat-bed tractor trailer or on a railroad flatcar. These containers are a safe, radiation-shielded mechanism for moving spent nuclear reactor fuel, and they have been tested for breakage in simulated collisions, train wrecks, and fires. Reactor fuel and other radioactive items have been transported across the country for decades without a single spillage accident.

Unfortunately, although the civilian nuclear power stations in the United States have been paying for a National Spent Fuel Repository for several decades, a licensed facility has yet to open. A nuclear power plant is designed and licensed to have sufficient storage volume for all the fuel it will ever use, so there is not yet a fear of having to shut down the nuclear power plants because of a glitch in the fuel transport plans. However, as the plants eventually outgrow their usefulness and are shut down, this fuel must be moved to a repository before a plant can be dismantled and the area returned to its original state.

SUITING UP FOR MAINTENANCE PROCEDURES

The weeks during a refueling operation are an excellent time to accomplish maintenance tasks in sections of the plant that are off limits during full-power operation. Areas near the reactor and the primary cooling loop are considered too radioactive for close hands-on work during power generation. It is possible to do these tasks during refueling or during a special maintenance shutdown, but the workers must be decked out in special clothing.

To work in such an environment one must have a known radiation-exposure baseline. Health effects of radiation exposure are considered cumulative, so a worker's past history of exposure must be known, and work time is strictly limited against the measured radiation-exposure rate. One worker is allowed a maximum total radiation exposure, a maximum yearly exposure, daily exposure, and hourly exposure. Records of each worker's exposure are scrupulously kept in the health physics office in the plant.

Radiation work also requires that a radiation suit be worn. A radiation suit offers no protection against radiation. It is not a shield, and it does not stop gamma rays, X-rays, or neutrons. It is designed for one purpose, to prevent contamination of the worker by radioactive dust. A worker is never directed to operate in a dangerous field of radiation, but when moving in an area where radioactive materials have been present, it is possible for a person to pick up radioactive dust. Wearing ordinary work clothing, there is no way to avoid it. Dust works its way into the palms, sticks to the hair, and is certainly breathed into the nasal passages and lungs. It even finds its way into the gastrointestinal system and works its way behind the eyes. Dust may not be at all dangerously radioactive, particularly if exposure is brief, but the effects are cumulative. Dust, lingering in or on a person, can do damage if allowed to remain in contact for a length of time. Radioactive dust can be difficult to detect, and it can be most persistent in clinging on and in a human body. Although any nuclear plant has aggressive shower facilities to clean a worker's body of contamination, it is best to never make skin or breath contact with such dust.

The radiation suit is a hermetic, airtight seal worn on the body, with breathable air supplied through HEPA, or high-efficiency particulate air, filters. The HEPA filter was invented for the Manhattan Project, the atomic bomb development effort during World War II. Although the design is now the world standard for all protective filters, it was designed specifically for radioactive dust. By definition, a HEPA filter removes at least 99.97 percent of airborne particles 0.00001 inches (0.3 μm) in diameter.

Donning a radiation suit begins with the overalls. Depending on the type of work to be done, these can be made of anything from a light, disposable material to heavy rubberized cloth. The garment is one piece and zips in the front, with long sleeves and pant legs to the floor. One size does not fit all, and the suit must fit snugly but allow full movement. It is made to be tear resistant, so that no air inlet is likely to be formed by snagging it on projecting hardware.

Over the feet go socks of the same material, with elastic openings to fit tightly on the ankles. Over the socks go seamless plastic boots, and the transition from boot to pants leg is sealed with duct tape. Next go the gloves, and these too are sealed at the wrist with duct tape.

Over the head goes the mask, completely sealing the head against dust intrusion to the mouth, eyes, nose, or hair. Having a beard is discouraged, as it can prevent a seal on the lower face. This air seal at the nose and mouth is important, even if the head is sealed against dust by the large collar. The wearer must be able to create a hard suction against the HEPA filer mechanism. A round HEPA canister sticks out on both sides of the face, and plastic goggles are over the eyes.

Wearing the radiation suit, a worker can perform hands-on tasks with equipment in a normally off-limits section of the plant without fear of unhealthy contamination. Workers can dismantle valves, repair leaking pipes, replace filters, or do anything that requires hands-on maintenance with no fear of accumulating radioactive dust. In addition to the anti-contamination suit, each worker is supplied with an integrating dosimeter to monitor his total radiation exposure from external sources. At least one worker on a team is a health physicist who carries a battery-powered radiation monitor. This instrument gives a moment-to-moment reading of the rate of radiation exposure to the workers. Removing the suit is a ritual designed to prevent one from smearing any dust contamination off the outside surfaces of the suit and defeating its purpose. A suit having detectable contamination may be discarded as a low-level waste item after a single use.

This has been a tour of a nuclear power generating station, from the front gate to the refueling floor, giving a brief glimpse at the culture and the physical plant of this special industry. The future of this power source may depend on its economic advantage over other, competing energy industries, such as solar collection or wind energy. The next chapter will give some guides as to what to expect in the next few decades, as the power generation of the world becomes a dynamic force of commerce, comfort, and human survival.

8 Environmental Advantages, Disadvantages, and Economics

In October 1927, the Ford Motor Company was ready to unveil a secret project that had consumed all aspects of this automobile-manufacturing complex for the past year. Intense, round-the-clock engineering had gone into developing a new car from scratch, to replace the now obsolete Model T Ford that the company had built for the past 18 years. No expense had been spared, and the team of expert engineers had thought of everything, from the all-weather tread pattern on the tires to the rain gutters on the top. There was only one feature of the new car that they had not worked out in detail. They had, in fact, never even thought of it. For what price was this car going to sell? They were designing the sticker to go in the side window, and they had not a clue as to what the bottom number should be.

They turned all eyes to the patriarch of the organization, the man who had founded the company and made it the leader of the automotive world, Mr. Henry Ford. He briefly narrowed his eyes at the question and asked one back, "How much does it weigh?"

The head of engineering, somewhat baffled by the counterquestion, knew not to hesitate in answering it. He quickly turned to the last page in his specification list and read out the answer, "Two thousand, two hundred sixty-five pounds."

Ford produced a folded envelope from one pocket and a stub of a pencil from another. He scribbled a few seconds, looked up, and gave his answer, "Five hundred twenty-five dollars, FOB Detroit."

Henry Ford, as it turned out, was quite familiar with the process of building a car, starting with iron ore, a forest of trees, and a rubber plantation. He knew how much it cost to turn a pound of material into a car. He multiplied it by the weight in pounds and added 10 percent profit. If only it were that easy to calculate the cost of energy. Not only is it difficult to calculate the cost of electrical generation, it is almost impossible to get a straight answer to this important question, particularly when one is comparing different generation modes, such as nuclear, coal, gas, wind, hydro, geothermal, or solar. There are hidden costs, government subsidies, taxes, lifetime issues, and efficiencies to be considered. Some energy technologies are simply too new to have produced answers to the important questions. Nuclear power, while it has been generating electricity for the past half century, is still in the experimental phase. There is no definitive knowledge of how long a nuclear plant should last or how much it should cost to tear one down, simply because so few billion-watt plants have reached the end of life.

The problem of calculating the cost per kilowatt of electricity is analogous to the old problem of computing the cost of transportation. In the 19th century, it would have seemed easy to find the cost of a train ticket from Atlanta to Baltimore. It was not. The cost was not simply dependent on the amount of fuel burned in a locomotive for 632 miles (1,017 km), the depreciation on the locomotive, the rolling stock, and the rails, plus the pay for the engineer, the stoker, and the conductor. It was much more complicated. The cost of putting a train in Atlanta, ready to pick up passengers and freight, was much different from the cost of putting the same train in Pocatello, Idaho. Some trains went great distances with very little paying cargo and some went short distances fully packed. Computing the cost of a train trip turned out to be an immensely complex nonlinear problem that would strain theoretical mathematics, and the definitive solution has never actually been found. Today, airlines have taken the place of railroads, with similar problems of costing. Stiff competition among airlines, wildly varying loading factors, and bouncing costs of fuel make it difficult to price a simple airline ticket.

In similar fashion, the cost of electricity depends on how much is demanded, which can depend on the weather, the time of day, the way it is generated, the distance it is transmitted, and the season of the year. The cost of generation depends on the spot price of fuel, if fuel is being used, or the percent of cloud cover, the wind speed, or the yearly rainfall. It costs less to send electricity to a consumer 10 miles from the generating station than it does to send it 200 miles, and yet the short-run customer is charged the same rate as the long-run customer, with the short-run consumer essentially paying a subsidy for the long-run. The problem of estimating the long-term cost of one type of power generation technology and comparing it with another can become difficult. As it turns out, advocates of a particular technology can usually find a mathematical method and data that will show that their chosen mode of generation can beat all other modes in cost of power production. The true bottom-line cost of power production, applicable to all production methods on a level playing field, has yet to be found. There are, however, some important factors that should be considered when pondering the cost of nuclear power.

AIR POLLUTION

A primary argument for nuclear power production as opposed to coal or natural gas is that a nuclear reactor produces no exhaust gas. In the United States, about 20 percent of the electrical power is nuclear and the rest is mainly coal and natural gas. Burning coal, natural gas, or any hydrocarbon fuel produces a heavy exhaust consisting of water vapor, carbon dioxide, trace gases released from the fuel, and partially burned fuel, or soot. The water vapor is harmless, the soot can be filtered out of the exhaust with enough effort, but the carbon dioxide escapes to the atmosphere. As the world population increases and the electricity demand of each human increases, the amount of electrical power demanded has increased, to the point where the carbon dioxide from power production may have become a serious contaminant in the biosphere.

Carbon dioxide is a normal component of the Earth's atmosphere. Plant life depends on it. Plants soak it up, add water, and produce structural sugars, such as cellulose, using the photosynthesis process, which allows them to grow and multiply. The exhaust from photosynthesis is oxygen, which is needed by animal life, including human beings. Animal life returns the favor by exhausting carbon dioxide back into the atmosphere. The presence of carbon dioxide is thus essential to the cycle of life and the well-being of Earth. The normal carbon dioxide content of the atmosphere is quite low, at only 0.038 percent.

Technically, all carbon-based, organic objects on the Earth's surface will eventually revert back to water and carbon dioxide, from which they were originally built, or at least as long as there is free oxygen in the atmosphere. This includes the reader, the wooden desk at which the reader may be sitting, the chair under the reader, and possibly the floor on which the chair is sitting. All hydrocarbon-based objects, including logs, lumps of coal, or cans of gasoline, will give up energy when oxidized or burned in the reversion process to original components. This derived energy is that which was soaked up from the Sun by the plant that built the original hydrocarbon that eventually became fuel or food. Burning hydrocarbon fuel in the free oxygen of the atmosphere has been a common mode of energy production since the discovery of fire.

The problem of introducing carbon dioxide back into the atmosphere by burning has become problematic since the demand for derived energy has ballooned as populations, transportation systems, and needs for electricity have increased. We can no longer satisfy the power demand by simply burning trees. Now, we need more power than all the trees in the world can provide, and we find the fuel for this increased production hidden underground. Coal mining and oil drilling have made up for the energy shortfall, allowing us to generate power at a rate that meets the larger needs for electricity. These mined resources are deferred hydrocarbon sources, or burnable assets that were not in the inventory of carbon on the Earth's surface. This relatively new, large source of carbon dioxide released into the atmosphere may be overloading the system, putting more carbon dioxide into the delicate balance of atmospheric gases than can be handled by the plant-animal cycle. Studies indicate that the percentage of carbon dioxide in the atmosphere has been rising rapidly for the past 200 years, or ever since we came to use coal and oil as fuel.

The rise of the atmosphere fraction has been sharp, but still the percentage of carbon dioxide in the atmosphere is slight and is very far from poisonous levels. However, at even slight concentrations, carbon dioxide has been shown to trap heat on the surface of the Earth. Regardless of how small this effect may seem, the chaotic nature of world climates and weather patterns is sensitive to change. It is difficult to predict how, but global climate will be affected by this shift in atmospheric chemistry.

A certain way to remedy this situation is to treat coal and oil combustion as a temporary, stop-gap mode of energy production that served us well until we developed a more permanent source of power, controlled nuclear fission. This newly perfected electrical generation technique can take the place of coal- and gas-burning plants as they reach their ends of life, and a combination of nuclear and other renewable sources can meet the ever-increasing energy needs of society. Using a fuel-breeding program, the world can essentially run on nuclear energy as a base power source forever, without straining the capacity of the biosphere to cycle carbon dioxide.

Although a large part of the carbon-dioxide excess comes from personal transportation vehicles, even this may change. In the next century, electric cars may take the place of gasoline-powered automobiles and trucks. This will add to the global electrical load as it decreases the amount of burned hydrocarbon fuel. This shift from a polluting energy use to a nonpolluting energy use may benefit the entire planet.

In 2005, 15 percent of the world's electricity was made by nuclear production, with most of the rest derived by burning coal and natural gas. This fraction dropped to 14 percent on July 16, 2007, when an earthquake in western Japan shut down all seven reactors at the Kashiwazaki-Kariwa Nuclear Power Plant. Still, the United States, France, and Japan account for 56.5 percent of nuclear-generated electricity. The European Union derives 30 percent of its power from nuclear sources.

Steam and fumes emerge from a coal-fired power plant in Niederaussem near Bergheim, Germany. *(Oliver Berg/dpa/landov)*

When discussing the relative dangers of gas emissions of coal-based energy production versus that of nuclear-based generation, it is important to know that there are controlled, periodic gaseous emissions from any nuclear power plant. Nuclear fission produces radioactive gases as well as radioactive solids. All can be chemically trapped and kept out of the atmosphere except the inert gases, xenon-135 and krypton-85. These gases

AN ODD REACTOR EXPERIMENT IN GEORGIA

Any power reactor in the world is heavily shielded against the release of radiation, and plants built in the United States are covered with multiple layers of protective armor to prevent a radioactive accident even if the worst possible breakdown occurs in the plant or if an airliner crashes into the reactor building. These extreme safety measures are part of the economic difference between nuclear power production and the more common coal-fired plants. The multiple layers of shielding are expensive but considered necessary. In the most severe nuclear accident to happen in this country, the Three Mile Island reactor core meltdown in 1979 near Harrisburg, Pennsylvania, the armor held and no radiation escaped into the surrounding area, even though the Unit-2 power plant was a total loss.

The question of what would happen if there were no radiation shielding nor mechanical armoring around a nuclear power reactor has been answered in a little-known series of experiments performed half a century ago in a secluded forest near Dawsonville, Georgia.

In 1957, the U.S. Air Force was in the middle of a secret project to develop a nuclear-powered strategic bomber, designed to remain in the air for months at a time using nuclear fission. Being an airplane, this weapons platform could have none of the usual concrete, steel, or lead shielding used in any other nuclear reactor installation. The weight had to be stripped down to a minimum and very little shielding was planned. As the program began tests of fission-powered jet engines, there were several unanswered questions concerning the effects of these naked reactors on plants, animals, and people. Would a nuclear bomber cut a swath of destruction on the ground as it flew over and what was the minimum amount of radiation shielding necessary for the flight crew on board?

To answer such questions definitively, an extensive laboratory was built in a sparsely populated forest, 50 miles (80 km) north of Atlanta, Georgia. It was assigned react chemically with nothing and are therefore difficult to store as waste. However, the more active gas, xenon-135, has a short, nine-hour half-life, and it is a β- emitter. This fission product makes its way easily from the fuel into the primary coolant water and is stripped out in a degassing operation in the cooling loop and compressed into a tank. After a wait of 90 hours it has passed through 10 half-lives and is considered to be of low-

the name AFP-67, or the Georgia Nuclear Aircraft Laboratory. The grounds were protected from intrusion by a high chain-link fence and a force of security guards, rotating duty 24 hours a day. At the geometric center of the rectangular, 10,000 acre (40 km^2) property was the test facility, a 10-million-watt nuclear reactor, completely free of any shielding and held 10 feet above ground by a hydraulic piston. The reactor was covered by a tin building and could be lowered into a 30-foot (9-m) pit in the ground to cool off after a test. It was dangerous to be anywhere near it when it was powered up, so the two-story control room complex had to be located 50 feet (15 m) underground.

Results of testing were predictably shocking. Every living thing within 1,000 feet (305 m) of a full-power test died, including viruses, bacteria, trees, grass, and any insect or bird that happened to be flying through. Live test subjects were brought near the reactor on special rail cars for carefully observed survival trials. Neutrons from the reactor, free of any restraint to diffuse away in the air, activated argon-40 in the atmosphere into radioactive argon-41. Neutrons penetrated the ground around the reactor and woke up some little-noticed trace elements in the Georgia clay. Europium-153 became radioactive europium-154.

With results of this testing it became clear that while it was possible to build aircraft engines that run on fission, it was simply too dangerous for serious implementation. The aircraft nuclear propulsion program was shut down in 1960, but the Georgia Nuclear Aircraft Laboratory continued with unshielded reactor testing until 1970. Today the laboratory grounds are a nature park, with a few immovable concrete remnants of the former laboratory remaining. The europium-154 in the ground around the reactor still lights up a *Geiger counter* with faint gamma rays, but nature has taken over the once naked dead ground. Within hours after a test, crabgrass would begin to germinate in the sterilized dirt, and now trees, weeds of all descriptions, and dense brambles grow through the foundation of the reactor building. Biology seems to have taken back the circle of death surrounding the deactivated reactor.

enough activity to be released into the atmosphere. The release of inert, radioactive gases, such as xenon-135 and krypton-85, is allowed under the federal rules for safe operation of the plant.

The inert radioactive gases have a practically zero body burden, meaning that there is no biological use for inert gases, and they cannot bind chemically to any tissues. There is essentially no danger from the radioactive gases emitted by a nuclear power plant. For it to affect animate life, radioactive contamination must be bound mechanically or chemically to the body, so that the effects can accumulate over a long period of time. This unique type of contamination, from the inert gases, quickly dilutes to undetectable levels in the atmosphere.

There is a larger degree of radiological danger from the operation of a coal-fired power plant than from a similarly sized nuclear plant. The typical lump of coal contains 1.3 parts per million uranium and 3.2 parts per million thorium, both of which are radioactive elements. A typical coal-fired plant releases 5.2 tons (4.7 mt) of uranium and 12.8 tons (11.5 mt) of thorium per year into the atmosphere as uncaptured fly-ash and into the ground and water environments as cinders. In the United States, the yearly population exposure to radiation from continuous operation of a billion-watt nuclear power plant is 4.8 person-rem (0.048 person-Sv). From continuous operation of a billion-watt coal-fired power plant the yearly population exposure is more than 10 times that, or 490 person-rem (4.9 person-Sv).

THERMAL POLLUTION

All energy conversion systems in which an available form of energy is converted into usable electricity have some effect on the environment. These man-made systems invariably change some long-established condition of the local ecosystem. Windmills stop wind from blowing naturally over the landscape, solar collector stations prevent sunlight from hitting the ground, and hydroelectric dams halt flowing rivers. In the case of any mechanism that converts heat into electricity, such as a coal, natural gas, geothermal, or nuclear plant, excess heat must be shed into the environment. This special form of environmental intrusion is thermal pollution.

A highly efficient heat engine such as a multistage turbine running under optimized conditions can convert only about 40 percent of the available thermal energy into electrical power, and the remaining 60 percent must be dumped back into the surrounding territory. Flowing water is a very efficient coolant and is used in most high-output nuclear, coal, and gas plants. Coal plants are built with the goal of using as little coal as possible, so the turbogenerator machinery is designed for as much efficiency as can be achieved. The turbines in these plants are unusually complicated and require a lot of tender care to keep them running. Nuclear plants are built with a different goal, to achieve safety and reliability at the cost of everything else. The cost of fuel is not really a consideration, and it is best to keep the machinery on the turbine deck simple and robust. As a result, a nuclear plant runs at about 33 percent efficiency, while a coal-fired plant runs at 40 percent. Seven percent more heat is discharged into the flowing water at a nuclear power plant than at a coal-fired plant of a similar electrical output.

Some water used in a power plant cooling tower simply evaporates and is often visible as steam flowing out the top of the tower, but most of it is released back into the source, elevated in temperature 16°–36°F (9°–20°C). An example of a large nuclear generating station is the San Onofre plant, on the coast of California between Los Angeles and San Diego. Its two reactors generate a total of 2.2 billion watts of electricity. To cool its turbine condensers, 2.4 billion gallons (9.0 billion l) of water per day are drawn from the Pacific Ocean. A flow rate of 830,000 gallons (3,100,000 l) per minute is maintained for each reactor unit. Water is sucked into the system through two intake structures located 3,000 feet (900 m) offshore, and the heated water is reintroduced into the ocean through a series of 63 exit pipes spread out over a distance of 2,450 feet (747 m). The water is heated by 19°F (34°C), but it quickly mixes with the ambient seawater. If San Onofre were a coal-fired plant, the water would probably be heated by 18°F (32°C). The average temperature rise in the ocean near this plant is less than 2°F (1°C).

Most aquatic organisms can tolerate only a small change in water temperature, so the thermal discharge from a power plant can cause havoc. Heated water contains less oxygen, plus organic material decomposes more rapidly in warmer water, and this further decreases the oxygen content. The exact effect of warming the water is somewhat unpredictable. In 1988, the Maanshan Nuclear Power Plant began operation on Nanwan Bay in Taiwan. Within days, the coral in the vicinity of the coolant discharge channel began to die, bleaching out and turning white. However, two years later the coral began to recover. It had apparently developed a tolerance to the higher temperature, after being hit with the shock of a sudden environmental change.

Water organisms are just as likely to benefit from warm water at a power-plant discharge point. In Florida, manatees, large marine mammals, may have been saved from extinction by the warm water coming out of power plants. They cannot survive in cold water, and the warmth of Florida rivers has been altered by increased agriculture that diverts natural springwater. When cold snaps occur in Florida, the manatees gather at the power station outlets and seem to enjoy the heated water. The same is true of sea turtles in California, the only large population of which is found near the discharge of a power plant in San Diego Bay. The intake and discharge pipes at many plants must be periodically cleansed of clogging organisms such as mussels that grow rapidly, seeming to thrive in the newly established environment.

FUEL AVAILABILITY

The typical nuclear power plant, with a generating capacity of a billion watts of electricity, requires about 440,000 pounds (200 mt) of mined uranium per year. The United States presently has 104 operating reactors demanding at least 49 million pounds (22,000 mt) of uranium per year as fuel. The world demand for uranium is more than 150 million pounds (68,000 mt) per year. World population is increasing along with the demand for electricity per person. The number of nuclear plants is expected to grow in the next few decades, and the demand for uranium will grow as well. By the year 2025, the uranium needed to fuel these reactors could grow to 220 million pounds (100,000 mt) per year. There is a developing question as to how long uranium fuel will be available.

Uranium is mined from the ground. Although it is a widespread element present throughout the Earth's crust, it is not common, and most deposits are in concentrations that are too slight for economical extraction. Uranium is not renewable. All the uranium that will ever exist is already here, none is being made, and that which is here is slowly decaying away into lead.

In 1956, M. King Hubbert (1903–89), a geophysicist working for Shell Oil, formulated the Hubbert peak theory. This theory states that the production of a fixed resource, such as uranium, will increase as demand increases. As demand increases, production follows, until the point where the resource demand equals the maximum rate at which the resource can be supplied. This will be the peak of supply. Beyond the peak, demand increases, but the supply rate goes down, as sources run dry. The price goes up, but the resource still eventually dries up. This theory was originally applied to oil, and it has been proven, as the source of oil has been outstripped by the demand for it. The price of gasoline has definitely increased as the world experiences the downside of the oil peak. On the leading side of the peak, it is a buyer's market. On the trailing side of the peak, it is a seller's market. There is no doubt that uranium, being a limited, nonrenewable resource, will experience a similar peaking. The only question is when this peak will occur.

Although the greatest number of nuclear power plants is currently in the United States, the demand for fuel may be driven by China and India, both of which are planning massive nuclear power expansions. China plans to build 32 nuclear power plants by 2020 with 40 billion watts of electrical capacity by 2020. By 2050, 300 or more nuclear plants may be added to China, if economical growth predictions hold true. India plans to add 20 billion watts of nuclear power by 2020.

Although this increase in demand for a limited resource may seem alarming, the uranium peak phenomenon is not nearly as serious as the peak for oil production, which is expected to occur early in this century. Using known sources of uranium in a growing world energy economy and with no fuel reprocessing, the supply of uranium fuel should be exhausted in 85 years. However, most of the Earth's dry surface has yet to be explored for economically practical uranium sources. Theoretical projections, given these undiscovered uranium mines, extend the burnout by 270 years. The eventual use of fast breeder reactors extends this very conservative estimate to 8,500 years.

Estimates of the peak production date for uranium do not take into account unconventional sources for this essential fuel. Uranium oxide, the mineral form of uranium, is soluble in water. This means that most of the uranium on Earth is not locked in mineral deposits. It is dissolved in the oceans. Seawater is estimated to contain 8.8 trillion pounds (4 billion mt) of uranium. At the current rate of consumption, there is enough fuel in the oceans alone to power the world's nuclear reactors for 58,000 years. Although the concentration of uranium in seawater is very low, 0.003 parts per million, proven extraction methods are passive and require no digging, rock-crushing, or chemical processing. The best high-grade uranium ore, such as that found in Canada, is 20 percent uranium, or 200,000 parts per million, and the ease of processing such rich ore has kept the price down. When the demand for uranium oxide reaches the point where the cost per unit weight is sufficient, or $300 per kilogram,

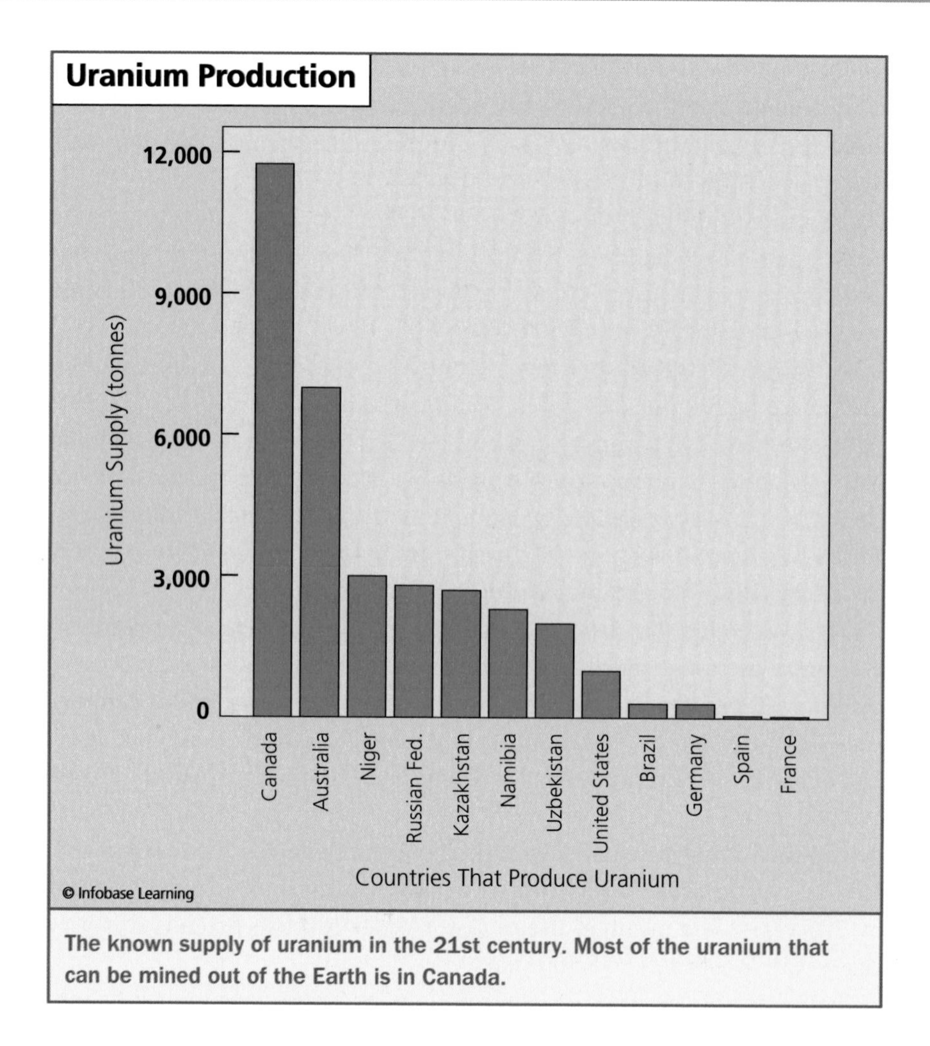

The known supply of uranium in the 21st century. Most of the uranium that can be mined out of the Earth is in Canada.

the water-extraction method of mining uranium will be activated. The current price rises and falls, as does the price of most commodities, and is currently around $80 per kilogram.

Another unconventional source of uranium is phosphate mine tailings. There are an estimated 44 billion pounds (20 million mt) of uranium locked in phosphate deposits. If phosphates are mined and processed for fertilizer at the current rate, extraction of uranium oxide from the leftover material would result in an annual output of 8.2 million pounds (3,700 metric tons). The technology for this extraction process has been developed and used. The cost is between $48 and $119 per kilogram of uranium oxide. There are also sources of uranium in coal ash. A single pile of waste from a coal-fired power plant in Xiaolongtang, China, contains an estimated 4.6 million pounds (2,085 mt) of uranium oxide. Oil shale also contains uranium as a by-product of processing, and uranium oxide has been extracted from this material in Sweden and Estonia.

The fuel burnout predictions cited here assume that uranium or plutonium bred from uranium is the material to be used for nuclear power. There is another reactor fuel, thorium-232. Although thorium-232 is not fissile, it readily converts into uranium-233, which is fissile under neutron bombardment. Experimental reactors have been successfully run using nothing but thorium-232 as fuel, and there is gathering interest in this option. Thorium is three times as abundant in the Earth's crust as uranium. Furthermore, natural thorium is 100 percent usable fuel, thorium-232, whereas natural uranium is only 0.72 percent usable uranium-235. There is no need for enrichment with thorium fuel, and there is no waste. Current experiments in thorium-fueled reactors are under way in Canada and India.

With the uranium-oxide production rate at its current level, 30 percent of the world's known in-the-ground reserves are in Australia, at an estimated 2.52 billion pounds (1.142 million mt) of uranium. Kazakhstan has about 15 percent of the reserves, with 1.8 billion pounds (817,000 mt), and Canada has 980 million pounds (444,000 mt), or 12 percent. The world will not run out of uranium soon.

FUEL REPROCESSING

Currently in the United States, civilian nuclear plant fuel is loaded into a reactor vessel, used until the uranium-235 content of the uranium has dropped to about 1 percent, and pulled out to be put in storage. This is a simple process, but it does not take advantage of the remaining uranium-235 left in the spent fuel, nor does it make use of the plutonium-239 fissile fuel that has been converted from uranium-238 in the fuel. Raw, mined uranium is too inexpensive to warrant the cost of leftover fuel extraction from spent material.

The reprocessing of spent reactor fuel has an additional advantage over leaving it in the raw state. Only about 5 percent of spent fuel is highly radioactive fission products. By chemically removing the uranium and the plutonium from spent fuel, the volume of waste that must be stored is greatly decreased. This simplifies the transportation of waste material, which must be moved in heavy, shielded containers, and it effectively

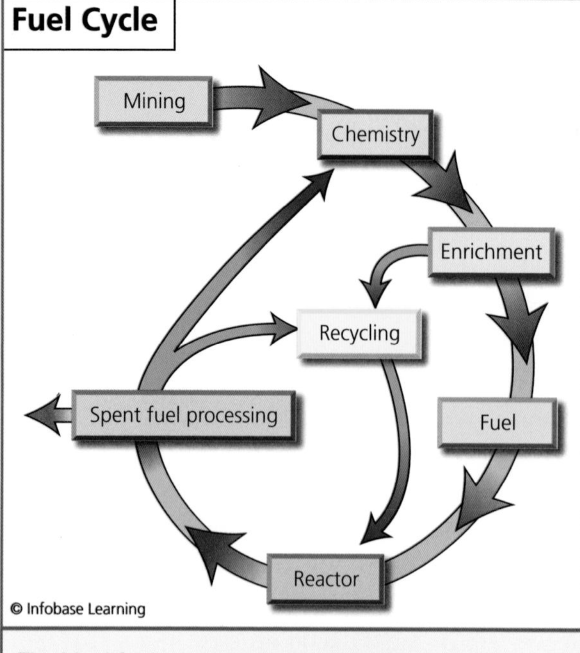

The ideal fuel reprocessing cycle. Instead of the wasteful practice of burying spent fuel, it can be reprocessed to gain as much energy out of the uranium as possible.

increases the capacity of a storage facility. Useful medical and industrial isotopes are removed from reactor waste in the reprocessing, and sale of these commodities plus the sale of the extracted fuel can defray at least part of the cost of the process.

Although fuel reprocessing was invented and perfected in the United States beginning in 1943, commercial spent reactor fuel is only reprocessed in Europe and Asia. The two large-scale fuel-reprocessing plants are at Sellafield, England, with a recovery capacity of 2.6 million pounds (1,200 mt) of uranium per year, and COGEMA La Hague, France, recovering 3.5 million pounds (1,600 mt) per year. Smaller-scale plants operate at the Mayak Chemical Combine in Russia, the Tokai Plant in Japan, and the Tarapur Plant in India. The last commercial reprocessing plant in the United States was at West Valley, New York. It was closed in 1972, when changing regulatory requirements reduced its profitability to below the break-even point. Large government-owned reprocessing plants at Hanford, Washington, and Savannah River, South Carolina, operated until the end of the cold war, extracting weapons-grade plutonium-239 from spent reactor fuel.

The current method of plutonium and uranium extraction from nuclear waste is the PUREX (plutonium and uranium extraction) process, developed at Oak Ridge, Tennessee, in 1949 under government contract. Although there are several known methods for fuel extraction, PUREX and its five modifications, UREX, TRUEX, DIAMEX, SANEX, and UNEX, are used worldwide for reprocessing.

In October 1976, President Gerald Ford issued a presidential directive suspending the commercial extraction of plutonium from used reactor fuel, after India built and demonstrated a nuclear weapon made with commercially reprocessed fuel. In April 1977, President Jimmy Carter

Inside the THORP nuclear fuel-reprocessing plant at Sellafield, Cumbria, England *(Photofusion Picture Library/Alamy)*

sealed the fate of commercial reprocessing in the United States by fully banning the handling of spent fuel. Although this ban was lifted in 1981 by President Ronald Reagan, the total loss of private venture in a new processing plant in Barnwell, South Carolina, after Carter's 1977 edict discouraged further investment. This situation will remain stable until the price of new, mined fuel rises to the point at which fuel from reprocessed nuclear waste becomes valuable. At this time, one large-scale reprocessing plant is under construction, at Zheleznogorsk, Russia.

THE COST OF NUCLEAR POWER

Electricity generated by nuclear fissian has the potential to be the answer to many problems of power generation and usage. National dependence on fossil fuels, such as oil, for transportation has caused global trade imbalances and outright wars with foreign sources. Further dependence on coal and natural gas for electrical power is beginning to affect air quality and even climate. A shift to nuclear power could solve these issues.

However, power generation and resource importation are money-making enterprises, and every aspect of them depends on the balance between profit and loss. At the earliest beginning of the national push for nuclear power as a clean, advanced way to generate power, this option was given many government subsidies and incentives for development. The first problem was a complete lack of domestic uranium sources. The United States did own a stockpile of uranium and plutonium, but these were strategic resources for weapons production that had been obtained from mining in Africa and Canada. As an incentive for uranium exploration and mining in the home territory, the Atomic Energy Commission (AEC) set an artificially high price for uranium ore. At the same time, it was made illegal to own uranium. Only the federal government could own uranium, so the government would buy and take possession of any extracted from a commercial mine.

As an incentive to generate nuclear power, the fuel was free. A nongovernment-owned reactor could only run on fuel borrowed from the AEC. This meant that the government kept absolute control of every pound of uranium in the United States, and it was artificially inexpensive to generate nuclear power. There was no uranium to be bought.

As a further incentive, the expensive development, design, and testing of key reactor components was paid for by the Department of Defense, under the auspices of the Navy nuclear submarine program. Early commercial plants would have been impossibly expensive without billions of dollars having been spent to perfect submarine and aircraft carrier power plants. Electrical power plant designers and builders had a range of proven, nuclear-grade pumps, valves, vessels, steam generators, and piping to choose from without the withering cost of research and development. With no-cost fuel and preengineered systems, commercial nuclear reactor technology was able to start and gain a credible foothold in the power industry.

There was, however, a snag. There was a potential for an explosion or fire in a civilian nuclear plant. Such disasters were not unheard of in the power industry, which tended to use superheated steam to turn turbogenerators. In the case of the traditional coal-fired plants, such an accident could destroy a power plant and kill several people, but an accident in a nuclear power plant involved unknown factors. If a nuclear reactor were to suffer a steam explosion, it could throw radioactive contamination over hundreds of square miles of territory. There was a potential for enormous liability, and industrial insurance companies and investors were unwilling to accept the unqualified costs of an accident with unknown probability.

Responding to this hindrance in the development of commercial nuclear power, the U.S. Congress passed a new federal law, the Price-Anderson Industries Indemnity Act, in 1957. This act established a no-fault insurance system in which the first $10 billion of liability is covered by the nuclear industry. Claims above $10 billion are covered in full by the federal government. This law was originally to be dropped in 1967, as it was assumed that by then the nuclear power industry would have a proven safety record and standard insurance policies could be written. By 1966, it was clear that private insurance companies were still unable to underwrite nuclear power, so the Price-Anderson Act was extended until 1976. Since then, the original act has been slightly amended and extended five times. It is presently scheduled to run out in 2025.

With the final hurdle eliminated by the Price-Anderson Act, commercial nuclear power was able to break into the U.S. economy and compete with other means of generating electricity. In the past 53 years, a total of $151 million has been paid from primary insurance to cover claims related to nuclear power. This includes $65 million paid by the *Department of Energy* to cover claims under liability for government-owned nuclear operations, and $70 million paid out for one reactor accident, the Three Mile Island reactor meltdown in 1979. As a comparison, the Tennessee Valley Authority Kingston Fossil Plant, a coal-fired generator in Roane County, Tennessee, experienced a coal ash dike rupture in a rainstorm on December 22, 2008. A mound of 1.1 billion gallons (4.2 million m^3) of coal ash covered 300 acres (1.2 km^2) of residential property, damaging and destroying homes. A pending lawsuit against the Tennessee Valley Authority for this single coal-power accident by residents tops $165 million. This one liability suit is greater than all nuclear power insurance payments in history. Since its passage in 1957, no money has been paid to claimants by the federal government under the Price-Anderson Act.

Still, the ultimate competition between nuclear power and other energy sources may depend on two factors, the cost of building the plant and the cost of running the plant. Not including the cost of the land on which a power plant is built, the cost of a new coal-fired, 1-billion-watt generating station is about $1.2 billion. A nuclear plant of similar generating capacity costs about $3.2 billion. The nuclear plant is more complicated, with more stringent safety factors, and is simply more expensive to construct.

The cost of operation of a plant is a complex calculation, as it depends on everything from the availability of the generator to the lifespan of the facility. A generator is unavailable when the plant is down for maintenance or repairs or, in the case of a nuclear plant, is down for scheduled refueling. The availability of nuclear plants has improved over the past 40 years and now tops 90 percent, making it even more reliable than coal-fired power. A coal-fired or nuclear power plant can have a lifespan of up to 40 years.

A major difference in coal and nuclear operating costs is the fuel, for which a nuclear plant must now pay the international market price. Fuel for nuclear power is about five dollars per megawatt-hour, while coal is $11 per megawatt-hour.

Operating and maintenance costs are understandably higher for a nuclear plant, at six dollars per megawatt-hour compared with five dollars per megawatt hour for coal. Pensions, insurance, payroll taxes, property taxes, administration, general overhead, and capital costs are the same in both cases and total up to $13 per megawatt-hour. Regulatory fees for nuclear power are much higher than those for coal, at one dollar versus 10 cents per megawatt-hour. A large discrepancy is the cost of decommissioning and waste disposal. A nuclear plant can be licensed for construction only if there is a plan for deconstructing the plant, and the cost of this future activity is factored into the operating cost, at five dollars per megawatt-hour. A coal plant is built with no plan or set-aside money for tearing it down, and no cost is associated with disposing of the coal ashes. As a result, coal power costs about 90 cents less per megawatt-hour than nuclear. These costs, in terms of dollars per megawatt-hour of power produced, are compared in the following table.

This is a crude estimation of the compared cost of two methods of generating electrical power. The true costs depend on such considerations as environmental effects and variable fuel availability. A factor that could definitely tilt the comparison is the imposition of a federal carbon tax. It is possible that the government could charge a fee to industries, with the magnitude of the cost based on the amount of carbon dioxide emitted into the atmosphere. Under this condition, coal-fired power generation would sustain a crippling blow, and nuclear power, as well as other renewable energy sources, would suddenly become much more profitable than the traditional power source. The goal of such a legislated law would be to reduce the carbon dioxide load into the atmosphere.

A COMPARISON OF ITEM COSTS FOR NUCLEAR AND COAL POWER PLANTS OF SIMILAR SIZE			
Item	**Cost Element**	**Nuclear $/Mw-hr**	**Coal $/Mw-hr**
1	Fuel	5.00	11.00
2	Operating and maintenance—labor and materials	6.00	5.00
3	Pensions, insurance, taxes	1.00	1.00
4	Regulatory fees	1.00	0.10
5	Property taxes	2.00	2.00
6	Capital	9.00	9.00
7	Decommissioning and DOE waste costs	5.00	0.00
8	Administrative/overheads	1.00	1.00
Total		**30.00**	**29.10**

The coming decades hold promise for the continuation and even the expansion of the percentage of the world's electrical power generated by nuclear means. As the world population increases and the accompanying electrical load grows, all means of generating power will have to grow as well. A portion of that growth will be nuclear into the foreseeable future.

Conclusion

On March 11, 2011, the Tōhoku earthquake and tsunami on the east coast of Japan destroyed the Fukushima I Nuclear Power Plant. Of the six boiling water reactors at the plant, located on the beach at Ōkuma, the three General Electric units were producing power at the time, with the three Hitachi reactors off-line for maintenance. Although all three reactors made an orderly, automatic shutdown on the first indication of ground movement and sustained no significant damage, the earthquake had occurred 81 miles (130 km) offshore, and within minutes a 33-foot (10-m) wave of water crashed into the plant. The plant had been built to withstand earthquakes, and even this huge wave did not disturb the reactor building or its nearby turbine hall. It did, however, wipe away the power transmission lines and the switchyard.

Disconnected from all other power sources and completely shut down, the plant was left to its emergency generators to provide power to the water pumps that were necessary to keep the coolant flowing in the reactors, still hot from having run at full power just before the earthquake. The diesel generators, two per reactor, came on automatically, as designed. For all the emergency planning and all the preparations and fire drills, there was no plan for a 33-foot wave. The tanks of diesel fuel had been swamped. Water got into the fuel lines, and after one hour the diesels stopped running.

Having no power to run coolant pumps, all three reactors overheated. The coolant boiled out, and hydrogen gas was produced by the chemical reaction between steam and the overheated zirconium fuel cladding. Hydrogen collected in the reactor buildings, and two exploded when sparks set off the gas. On top of those problems, the spent fuel pools in the floors in all six reactor buildings began to overheat for lack of moving water.

In all, the Fukushima I Nuclear Power Plant was a disaster with wide complications, putting anyone within 50 miles (80 km) in range of fission product fallout. It is provisionally rated a 7 on the International Nuclear Event Scale (INES), putting it at the same level as Chernobyl. Workers close to the reactor were exposed to unusually high levels of radioactivity, and the agricultural region of northeastern Japan was contaminated with detectable radiation. A primary concern was the fission product iodine-131 finding its way into Japanese milk. Ingested iodine-131 is absorbed by the human thyroid, and because it is radioactive in significant doses it can induce cancer.

To put this concern into perspective, at the highest concentration of iodine-131 in Japan, to raise the probability of thyroid cancer by a measurable amount would require one to drink 58,000 glasses of milk. Iodine-131 has a half-life of only eight days. Eight days after the spent fuel pond in Fukushima reactor 4 overheated and started shedding iodine-131, the radiation from it had gone down by half and so had the risk. After 80 days, the level of environmental radiation from iodine-131 was undetectably low.

This incident is one of a series of nuclear power accidents that cause concern for safety in American-built power plants, particularly when an earthquake strikes. The ruined reactors in Japan, after all, were built in the United States more than 40 years ago. The nuclear plant in the United States most likely to be challenged by a temblor is the Indian Point Energy Center, located on the Hudson River 38 miles (61 km) north of New York City.

The maximum earthquake expected at the site is planned for in all nuclear plant construction projects, and this includes Indian Point. Although the quake that hit Fukushima was larger than what was planned, it was not the trembling ground that destroyed the plant, it was the wall of seawater that contaminated the diesel fuel supply. If we are worried about a large offshore earthquake and tsunami, then we should turn our attention to California, where the Diablo Canyon Power Plant, having two Westinghouse PWRs, is located on Avila Beach. Ten miles (ca. 3 m) offshore lies the Hosgri fault, which triggered a 7.1 magnitude quake on November 4, 1927.

Due to the threat of an earthquake from this fault, the Diablo Canyon plant was made to withstand a ground acceleration of 0.75 g (7.35 m/sec^2). Fukushima 1 was designed for a maximum ground acceleration of 0.18 g (1.76 m/sec^2), and it was not seriously damaged by a magnitude 9.0 earthquake. The danger of a tsunami is different from that of an earthquake, but waves, flooding, hurricanes, and tornados have all been planned for in California and at all nuclear power plants in the United States. At Diablo Canyon, the diesel fuel tanks are located underground, with no vents that can take on water. In case everything fails, the reactors can be cooled for many days using a man-made lake of freshwater, located in the hills above the plant. Seismically qualified pipes can conduct water to the reactors and the spent fuel pools using nothing but gravity.

Almost every reactor in the United States is located near water for use as an ultimate heat sink for cooling purposes. For this reason, flooding of a river, lake, or ocean is included in the plant design as a likely problem. At Indian Point and every other plant, the reactor and its auxiliary equipment is designed to operate in flood conditions. The Waterford Steam Electric Station, a nuclear power plant located 20 miles (32 km) west of New Orleans, Louisiana, had to shut down during Hurricane Katrina on August 28, 2005, as a precaution. As soon as the wind died down, Waterford resumed operation, providing power to an area that had suffered extreme flooding and devastation in the storm.

It has been the purpose of this volume to answer questions about generating electrical power using the well-established technology of the light-water nuclear reactor, from the basic principles of fission to the act

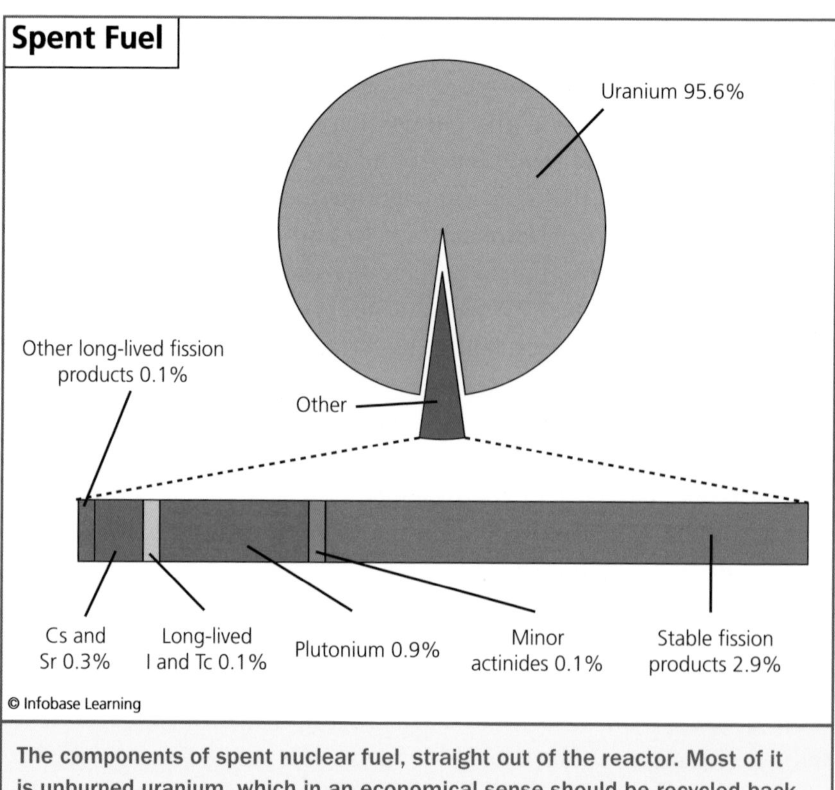

The components of spent nuclear fuel, straight out of the reactor. Most of it is unburned uranium, which in an economical sense should be recycled back into the power process either as fissile fuel or as breeder material.

of turning the mode switch on the power plant control console to the RUN position. While this commentary has touched on important issues of commercial power safety, the associated topic of waste disposal is also to be considered in the larger topic of nuclear power.

The task of safely transporting and storing radioactive fission products in spent reactor fuel was studied in a research effort funded by the National Academy of Sciences in 1957, just as the first commercial nuclear plant was being built in Shippingport, Pennsylvania. Previously, thousands of tons of radioactive high-level waste had been produced by some inefficient practices during the atomic bomb development efforts of World War II. This debris had been dumped in the ocean, liquefied and pumped into underground storage tanks, or temporarily stored in steel drums in out-of-the way places. These measures, while perhaps justifiable in the frantic atmosphere of a world war, would have to be improved for civilian power production. The National Academy of Sciences found that reactor debris should be consigned to underground storage and allowed to decay away into nonradioactive material over time. It was important that human beings should not be exposed to it before it had decayed to nondangerous levels of radiation, and a deep burial was the best place for it. Depending on the nature of the radioactive waste, which could involve complex decay cascades, storage for more than 1,000 years was advised.

There have been no improvements or modifications to this straightforward plan in the half century since it was formed, and engineering details have been worked out. An important consideration was that the geological formations in which radioactive waste is buried should not have water flowing through them. Flowing, underground water could eventually dissolve the material and transport it to wherever the water winds up, and a contamination of the water supply is to be avoided. An underground rock formation that has never had water flowing through it is a salt deposit. If water had ever been there, it would not be a salt deposit. The salt would have all dissolved quickly, and the formation would be a cave. For this reason, underground deposits of salt in deep bedrock were recommended for long-term nuclear waste storage sites. The National Academy of Sciences plan became known as the deep geological repository concept.

Although the plan was considered feasible and based on sound scientific judgment, the U.S. nuclear power industry and its federal regulatory and development agencies were slow to implement any form of this plan. For decades, nuclear plants were constructed with on-site temporary storage capacity for spent fuel, with no coordinated effort to consider the long-term consequences of a general conversion to nuclear power generation.

The first long-term underground storage for radioactive material was opened for experimentation in Germany in 1965. This government-owned facility was the Schacht Asse II, an old, played-out salt mine in the Wolfenbüttel District in Lower Saxony. After two years of study, the site was opened for fission and reactor fuel processing waste storage in 1967. By 1978, it was considered full, and no further material was stored. The facility has been monitored and studied since the closure, evaluating the feasibility of this means of storage.

Unfortunately, Schacht Asse II was excavated as a salt mine and was not originally intended to be a radioactive waste repository. There are no columns or struts supporting rooms, which are loaded with heavy, shielded concrete waste containers, and the mechanical stress is significant. Still, 125,787 drums of low-level radioactive waste with an activity of 49,000 *curies* (1.8×10^{15} Bq) are now in storage, plus 1,293 containers of medium-level radioactive material at 76,000 curies (2.8×10^{15} Bq). Most of the medium-level waste is from a fuel reprocessing plant in Karlsruhe, Germany, and specific regulations are in place to prevent a *critical mass* from forming as containers of uranium-235, uranium-233, and plutonium-239 are grouped together. No more than seven ounces (200 g) of uranium-235, for example, are permissible in a single drum.

Planning for the first underground storage facility in the United States began in 1974. An ideal site was identified in a 3,000-foot (1,000-m) thick salt formation, 2,150 feet (655 m) below the surface near Carlsbad, New Mexico. This geologic area had been stable and free of water for more than 250 million years. After 20 years of scientific study and testing, a full-scale underground storage facility was built, and it opened for operation in 1999. By 2006, the Waste Isolation Pilot Plant, or WIPP, had already successfully processed 5,000 shipments of radioactive waste. This plant, however, is owned by the U.S. government, and it is off-limits to commercial power waste. Only specific waste products from nuclear weapons development and manufacture can be stored here, but it is a well-managed test of the deep geological repository concept.

By 1982, the often-discussed problem of commercial nuclear waste storage was finally addressed by the U.S. Congress, with passage of the Nuclear Waste Policy Act. This piece of legislation established a time line and a procedure for establishing a permanent, underground repository for all the high-level radioactive material produced in commercial nuclear power plants. By July 1, 1989, 10 possible sites for two repositories, one in the east and one in the west, were to be recommended to the president of the United States. These recommendations were to include full environmental impact statements on each site.

Locations considered for the repository were the basalt formations at the Hanford Nuclear Reservation in Washington, volcanic tuff formations at the Nevada nuclear test site, and several salt formations in Utah, Texas, Louisiana, and Mississippi. Deep salt and granite formations in East Coast states from Maine to Georgia were surveyed. In 1987, the Nuclear Waste Policy Act was amended. The presidential approval of two out of 10 sites was eliminated, and one location, the nuclear weapon test area in Nye County, Nevada, was designated the repository. It is a ridge named *Yucca Mountain,* composed of volcanic tuff ejected from a now-extinct supervolcano. Money was assigned for study of the geological suitability of the site. Funds for the construction of the site were collected by combining yearly payments from all the nuclear power stations in the United States, with a promise of radioactive waste storage beginning by January 31, 1998. Additional funding was provided by the U.S. Navy, with a promise that nuclear ship and submarine waste could also be stored here.

Unfortunately, opposition to the use of this facility delayed the opening of the facility for more than a decade. In 2008, the Department of Energy began to slip the opening date to sometime in 2012–2015. In 2010, the Yucca Mountain Nuclear Waste Repository operating license was put in suspension, and its future as the only sanctioned location for spent commercial reactor fuel remains unknown.

No industry can remain productive if its waste material is allowed only to build up at the production site, and the lack of any fuel repository may be the most serious blockage to an otherwise viable method of the power generation. For the sake of future growth of the U.S. economy, and even the sustaining of its present activity level, this conundrum must be resolved.

Chronology

1942 On December 2, CP-1, the world's first nuclear reactor, goes critical and generates half a watt of power.

1946 In August, President Harry S. Truman signs the Atomic Energy Act, placing the nuclear energy industry under civilian control.

1947 In October, the Atomic Energy Commission begins work on investigations into peaceful uses of nuclear energy.

1953 The first boiling water reactor experiment, BORAX-I at INEL, proves that steam formation limits power and prevents runaways.

In December, the Atoms for Peace program is unveiled by President Dwight D. Eisenhower

1954 In August, declassified government documents are made available to civilian nuclear energy programs by a major amendment to the Atomic Energy Act of 1946.

1955 In January, the Atomic Energy Commission begins cooperation with civilian industry to develop nuclear power.

In July, BORAX-III provides Arco, Idaho, with power for a population of 1,200 for one hour.

1957 The International Atomic Energy Agency is formed with 18 member countries to promote peaceful uses of atomic energy.

In July, the first civilian power reactor, the Sodium Reactor Experiment, goes on line in Susana, California.

In September, President Dwight D. Eisenhower signs the Price-Anderson Act, protecting citizens, utilities, and contractors from lawsuits due to nuclear plant accidents.

In December, the first full-scale nuclear power plant in the United States (60 MW) goes on line at Shippingport, Pennsylvania.

1962 The first advanced gas-cooled reactor is built at Calder Hall, Great Britain, to power a naval vessel, for which it is found to be too big, and the reactor becomes a public utility for electrical power.

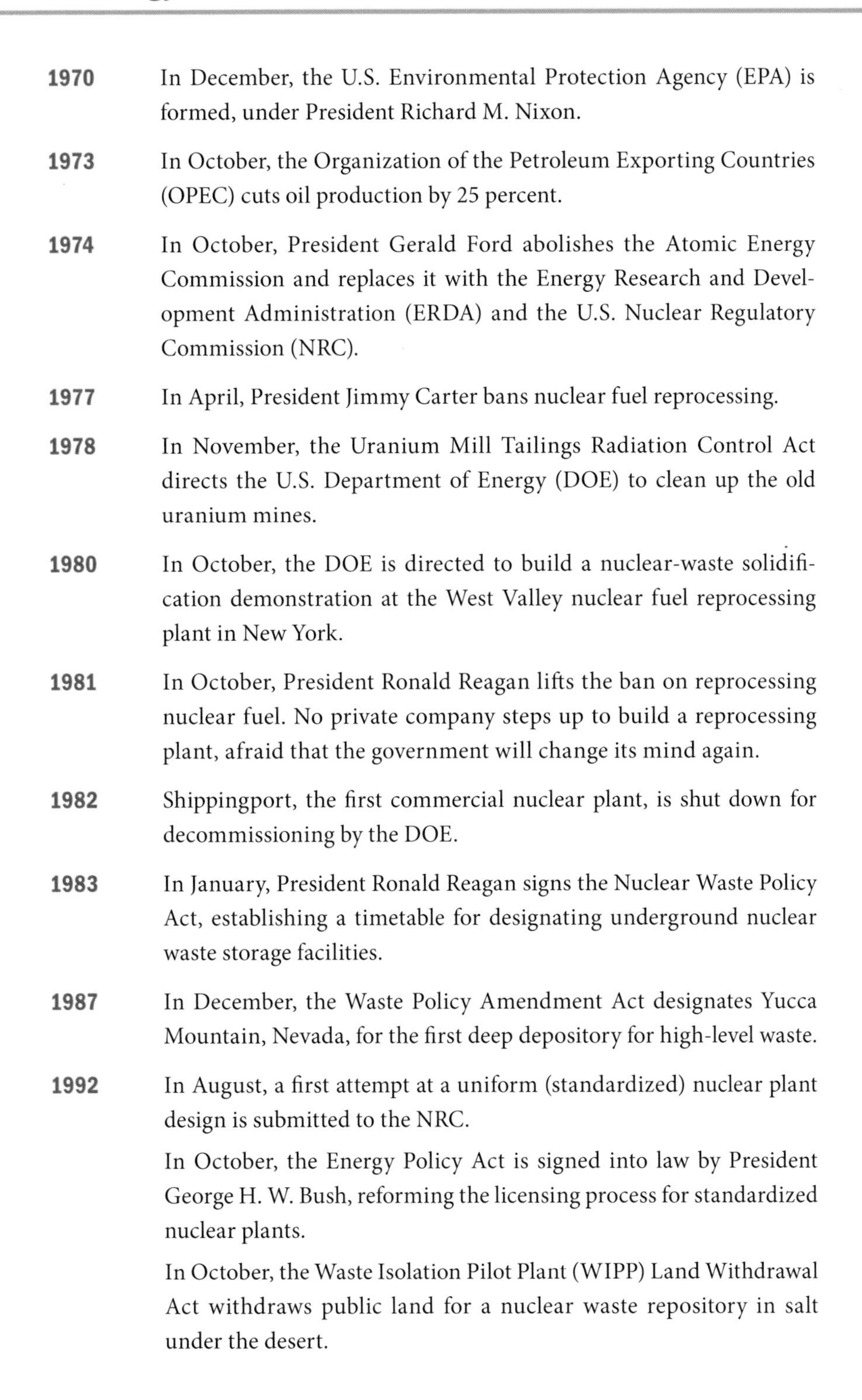

1970 In December, the U.S. Environmental Protection Agency (EPA) is formed, under President Richard M. Nixon.

1973 In October, the Organization of the Petroleum Exporting Countries (OPEC) cuts oil production by 25 percent.

1974 In October, President Gerald Ford abolishes the Atomic Energy Commission and replaces it with the Energy Research and Development Administration (ERDA) and the U.S. Nuclear Regulatory Commission (NRC).

1977 In April, President Jimmy Carter bans nuclear fuel reprocessing.

1978 In November, the Uranium Mill Tailings Radiation Control Act directs the U.S. Department of Energy (DOE) to clean up the old uranium mines.

1980 In October, the DOE is directed to build a nuclear-waste solidification demonstration at the West Valley nuclear fuel reprocessing plant in New York.

1981 In October, President Ronald Reagan lifts the ban on reprocessing nuclear fuel. No private company steps up to build a reprocessing plant, afraid that the government will change its mind again.

1982 Shippingport, the first commercial nuclear plant, is shut down for decommissioning by the DOE.

1983 In January, President Ronald Reagan signs the Nuclear Waste Policy Act, establishing a timetable for designating underground nuclear waste storage facilities.

1987 In December, the Waste Policy Amendment Act designates Yucca Mountain, Nevada, for the first deep depository for high-level waste.

1992 In August, a first attempt at a uniform (standardized) nuclear plant design is submitted to the NRC.

In October, the Energy Policy Act is signed into law by President George H. W. Bush, reforming the licensing process for standardized nuclear plants.

In October, the Waste Isolation Pilot Plant (WIPP) Land Withdrawal Act withdraws public land for a nuclear waste repository in salt under the desert.

1994 In January, Russian enriched U-235 is bought by the United States and down-blended for use in power plants, to keep it from being bought and used for bombs.

2000 In the United States, 110 nuclear power plants achieve a world's record for reliability, having operated at 90 percent capacity for the past 10 years, generating 2,024.6 billion-watt-years of electrical power without incident.

2009 On August 6, the NRC issues an early site permit and a limited work authorization to install two Westinghouse AP1000 Generation III+ PWR reactors at the Alvin W. Vogtle Electric Generating Plant near Waynesboro, Georgia. It is the first new reactor license issued in the United States for more than 30 years.

2011 On March 11, 2011, a magnitude 9.0 earthquake and tsunami lead to the destruction of the Fukushima I nuclear power plant on the northeast coast of Japan. The problems of scattered fission products over Honshu Island and the severe economic loss of the six-reactor plant cause reverberations throughout the international power industry. Planned expansions of nuclear power production are delayed.

Glossary

alpha particle also alpha ray, is a class of ionizing radiation composed of helium nuclei traveling at high speed. Alpha particles have a charge of +2, and are composed of two protons and two neutrons traveling stuck together. Alpha particles are highly energetic, but have little ability to penetrate anything. Alpha particles are emitted from heavy nuclei undergoing decay.

atom the smallest, most fundamental unit of an element, consisting of a central nucleus and a set group of orbiting electrons. The configuration of the electrons determines the chemical characteristics of the element.

atomic energy an antiquated term meaning energy that is released by the fission of heavy nuclei or the fusion of light nuclei. The better term is nuclear energy.

beta ray or beta particle, is either an electron or a positron ejected from a decaying nucleus. If it is an electron, then a neutron has decayed into a proton. If it is a positron, then a proton has decayed into a neutron.

boiling water reactor (BWR) A boiling water reactor is a commercial reactor in which light water is used as the moderator and the coolant. The moderator is allowed to boil in the reactor core, and the reactor uses no secondary coolant loop.

BORAX a series of experimental setups to test the concept of a boiling water reactor, near Arco, Idaho, at the National Reactor Test Station in the early 1950s

breeder reactor a nuclear reactor that makes more fuel, through neutron capture in non-fissile nuclei, than it uses to produce power. Breeders typically transform U-238 into Pu-239, and Pu-239 can be used as fuel.

Canada Deuterium Uranium (CANDU) a class of heavy water–moderated power reactors invented in Canada

chain reaction a series of chemical or nuclear reactions in which each reaction causes another reaction. A physical map of the reactions will show them connected, as if in a chain.

control rod or control, a metal rod made of a neutron-absorbing metal, such as cadmium, used to soak up excess neutrons in a nuclear reactor and bring it to perfect criticality by adjustment

criticality the balance state of a nuclear chain reaction, in which the number of neutrons being lost through leakage, unproductive capture, or fissioning capture exactly equals the number of neutrons being produced by fission

critical mass the effective mass of uranium or plutonium fuel at which a nuclear reactor is critical. The mass is effective because it can be artificially adjusted using neutron-absorbing controls.

curie a unit of measure of radioactivity. One curie is 3.7×10^{10} radiation-producing nuclear disintegrations per second.

decay a series of steps through which a radioactive isotope progresses, becoming different and successively lighter isotopes toward a final, stable nonradioactive isotope

Department of Energy (DOE) U.S. cabinet-level agency that regulates nuclear waste disposal and handling, nuclear weapons, and Navy nuclear reactors

deuterium heavy hydrogen. The deuterium is heavy because the nucleus contains both a neutron and a proton, and it weighs twice what an ordinary hydrogen nucleus weighs.

electron volt (eV) or the amount of energy required to raise one electron to a potential of 1 volt

element a pure chemical substance, consisting of one type of atom. An atom type is determined by the structure of its nucleus, and this configuration is unique to each of 118 elements known to exist.

enriched uranium uranium reactor fuel that has had the U-235 content improved. A typical power reactor uses fuel with the U-235 artificially increased to 8 percent.

enrichment the process of improving the concentration of fissile U-235 in natural uranium. Natural uranium, or uranium as mined, contains only 0.7 percent fissile U-235. The remainder is inert U-238.

fissile a descriptor for an element that will release energy and excess neutrons when fissioned

fission the splitting of a heavy nucleus into two lighter nuclei. Fission is caused by the absorption of a neutron in a fissionable element, and it can result in the release of excess energy.

fissionable a descriptor for an element that can be fissioned by neutron capture

fission products the lighter, always radioactive isotopes into which a fissile fuel breaks upon fission. Fission products are a wide range of isotopes, with half-lives from a few seconds to a few thousand years.

fuel reprocessing the chemical separation of various components of spent nuclear fuel. Reprocessing divides the spent fuel into radioactive fission products, unburnt fuel, and inert filler.

gamma ray a high-energy electromagnetic wave, above X-rays on the electromagnetic energy spectrum, originating in the nucleus. A gamma ray is generated when the nucleus experiences a rearrangement of subnuclear particles.

Geiger counter (Geiger-Mueller counter) an electronic radiation detector used to measure the presence of gamma or beta rays. The Geiger counter makes use of the extreme amplifying properties of an avalanche effect in a gas-filled tube excited by a high voltage.

half-life the time required for a radioactive sample to decrease its level of radioactivity by one half

heavy water deuterium oxide, or water made with two deuterium atoms and one oxygen in each molecule

isotope a sub-species of an element, distinguished by the number of neutrons in the nucleus. All possible isotopes of hydrogen, for example, have zero, one, or two neutrons in the nucleus, all of which have only one proton.

MAGNOX a now obsolete British reactor fuel formula (**MAG**nesium **N**on-**OX**idizing)

million electron volts (MeV) unit of energy applied to subatomic or subnuclear particles in motion

moderator any substance used in a nuclear reactor to slow high-speed neutrons from fission down to thermal speed. Neutrons at thermal speeds are more likely to initiate fission in a reactor.

neutron a particle of matter having no electrical charge. Neutrons are components of the nucleus in an atom.

Nuclear Regulatory Commission (NRC) the U.S. agency charged with the regulation and oversight of all nuclear activities in the United States

nuclide a subspecies of an element. The element is defined by the number of protons in a nucleus. The nuclide, or the isotope, of an element is defined by the number of neutrons in the nucleus. Nuclide is a finer distinction, because one isotope can have more than one nuclide, due to metastable forms.

nucleus (pl. nuclei) the massive center of an atom, built of protons and in all but one case, neutrons. Hydrogen is the only atomic nucleus having no neutrons, and only one proton.

plutonium element number 94 in the table of the elements. Plutonium is a chemically poisonous metal and is exceedingly rare in nature. Plutonium is commonly made by activating U-238, which decays to neptunium, which then decays to plutonium.

pressurized water reactor (PWR) The pressurized water reactor is a form of commercial power reactor using ordinary water as the moderator, held at high pressure so that it remains in the liquid state.

Pu-239 a fissile isotope of plutonium. One Pu-239 nucleus contains 94 protons and 145 neutrons.

Pu-240 a fissile isotope of plutonium. One Pu-240 nucleus contains 94 protons and 146 neutrons. Pu-240 is sensitive to spontaneous fission, not necessarily requiring a neutron capture to initiate the reaction.

radioactive capable of emitting radiation at a predictable rate by nuclear decay

radioactive decay the tendency of certain isotopes to undergo change in the nucleus. Any change in the nuclear structure causes radiation to be emitted from the nucleus. The time at which the change occurs is completely random and unpredictable, yet the rate at which a large sample of the particular isotope will decay is predictable and is characteristic of the isotope.

radioactivity the emission of radiation, either by the willful manipulation of a nucleus or by spontaneous nuclear decay

radiation a class of energy transmission by electromagnetic waves or by direct particle transfer. Radiation is invisible, and is undetectable with any human sense, but in extremely high dosage it can be very dangerous.

reactor a machine or system built to sustain a neutron chain reaction in a fissile material

thermal speed the speed at which air molecules move at room temperature as they bounce around and hit each other. At cold temperatures air molecules move more slowly, and at high temperatures they move more quickly.

tritium the heaviest isotope of hydrogen, having two neutrons in the nucleus. Tritium is slightly radioactive. It emits a low-energy beta particle.

U-235 a fissile isotope of uranium, having 92 protons and 143 neutrons in its nucleus. U-235 occurs rarely in nature, making up only 0.7 percent of the uranium found in the Earth's crust.

U-238 a non-fissile isotope of uranium, having 92 protons and 146 neutrons in its nucleus. U-238 makes up most of the uranium occurring in nature. It can decay indirectly into Pu-239 upon neutron capture.

uranium element number 92 in the table of the elements. Uranium is common in nature, and can be found in traces in most of the Earth's crust. It is mildly radioactive.

Yucca Mountain a deep geological repository storage facility for radioactive waste from power production located within a ridgeline on government-owned property in south-central Nevada

Further Resources

BOOKS

Cravens, Gwyneth. *Power to Save the World.* New York: Knopf, 2007. The author gives a very readable, personal account of the issues of expanding nuclear power in the 21st century.

Glasstone, Samuel. *Principles of Nuclear Reactor Engineering.* New York: D. Van Nostrand Company, Inc., 1955. Books that give further information on the design of nuclear power reactors exist only at the graduate level. This is the nearest thing to a readable account of nuclear engineering, but it still requires a knowledge of calculus.

Hore-Lacy, Ian. *Nuclear Energy in the 21st Century.* Maryland Heights, Mo.: Academic Press, 2006. This is an excellent reference covering the current designs and issues of nuclear power, with many useful tables, graphs, a glossary, and color photos, showing trends, projections, and a touch of history.

Mahaffey, James. *Atomic Awakening: A New Look at the History and Future of Nuclear Power.* New York: Pegasus, 2009. A history of nuclear power, showing how we got to the present situation, using newly declassified information and knowledge from inside the nuclear industry.

Marcus, Gail H. *Nuclear Firsts: Milestones on the Road to Nuclear Power Development.* La Grange Park, Ill.: American Nuclear Society, 2010. This book, available only from the American Nuclear Society, accurately cites first achievements on the long road to nuclear power, from the first controlled nuclear reaction in 1942 to the first nuclear power plant license renewal in the United States in 2009. It is full of surprises.

Tucker, William. *Terrestrial Energy: How Nuclear Energy Will Lead the Green Revolution and End America's Energy Odyssey.* Savage, Md.: Bartleby Press, 2008. A decidedly pro-nuclear argument for why nuclear power is a rational answer to growing needs worldwide for electrical power.

Zoellner, Tom. *Uranium: War, Energy, and the Rock That Shaped the World.* New York: Penguin Books, 2009. This is the sobering full story of the fuel that is used almost exclusively for nuclear power, from its ancient history to its future as a commodity to be fought over.

WEB SITES

Further depth on many topics covered in this book has become available on the World Wide Web. Although the Web is constantly expanding and changing and it would not be possible to cite a fraction of the important links to nuclear topics, the following list is a start.

American Nuclear Society (ANS) is a professional organization for nuclear scientists and engineers involved in research, development, and continuing operations of nuclear power plants in the United States. It publishes the magazine *Nuclear News,* as well as books and journals covering the nuclear disciplines. The public information portion of the Web site includes a section devoted exclusively to teachers and students, and education is a major concern of the ANS. Available online. URL: http://www.new.ans.org/. Accessed on December 15, 2010.

BWR Nuclear Reactor Technology Training has a free, downloadable boiling water reactor operator training program, complete with a reactor control panel simulator and a 108-page operator training manual. The simulator program gives you a realistic experience of starting a billion-watt power plant and bringing it up to full output. Look for the downloadable training manual in the training simulator. Available online. URL: www.acme-nuclear.com. Accessed on December 15, 2010.

Fukushima Daiichi Nuclear Disaster. There are many Web sites providing information on the Fukushima I reactor meltdowns and explosions, but the best and most complete account is located at Wikipedia. This write-up includes photos, diagrams, and clear explanations of the massive event that shook Japan and affected the nuclear industry in the rest of the world. Available online. URL: http://en.wikipedia.org/wiki/Fukushima_Daiichi_nuclear_disaster. Accessed on July 8, 2011.

Nuclear Street, the "nuclear powered portal," gives up-to-the-minute news of everything occurring in the nuclear industry. New books are reviewed, jobs in the nuclear industry are posted, and the current status of every nuclear power plant in the United States is updated in near real-time. Available online. URL: http://nuclearstreet.com/. Accessed on December 15, 2010.

Nuclear Tourist is an excellent, all-inclusive look at current issues and accomplishments of nuclear engineering all over the world, compiled and kept current by Joseph Gonyeau. Gonyeau is a highly experienced nuclear technologist, and he is quite familiar with the material he presents. This one site covers the entire discipline, from virtual tours inside a power plant to rules and regulations. Available online. URL: http://www.nucleartourist.com/. Accessed on December 15, 2010.

United States Department of Energy is a rich, complete site spanning the wide interests of this government agency, from energy science and technology to national security of energy sources. The Department of Energy owns, secures, and manages all the nuclear weapons in the military inventory. Although it concerns all sources of energy, nuclear power is a large component of the mission of the Department of Energy. This site is particularly accessible to students and educators. Available online. URL: http://www.doe.gov/. Accessed on December 15, 2010.

United States Nuclear Regulatory Commission is the most comprehensive government site for information concerning the control and safety regulation of the nuclear power industry. It provides a great depth of information concerning nuclear reactors, nuclear materials, radioactive waste, and nuclear security. There is a section that covers public meetings and involvement in nuclear power, as well as a constantly updated event report and news section. Available online. URL: http://www.nrc.gov/. Accessed on December 15, 2010.

Yucca Mountain Repository information is available at this official Eureka County, Nevada, Web site. All questions concerning the Yucca Mountain nuclear waste storage site, which remains very controversial, are answered in these pages. Find out why this mountain was chosen for the waste repository and how safe it may be. Available online. URL: http://www.yuccamountain.org/. Accessed on December 15, 2010.

Index

Italic page numbers indicate illustrations.